WATER STRUCTURE AT THE
WATER-POLYMER INTERFACE

WATER STRUCTURE
at the Water-Polymer Interface

Proceedings of a Symposium held on March 30 and April 1, 1971,
at the 161st National Meeting of the American Chemical Society

Edited by

H. H. G. Jellinek

Department of Chemistry
Clarkson College of Technology
Potsdam, New York

PLENUM PRESS • NEW YORK–LONDON • 1972

Library of Congress Catalog Card Number 70-189943

ISBN-13: 978-1-4615-8683-8   e-ISBN-13: 978-1-4615-8681-4

DOI: 10.1007/978-1-4615-8681-4

© 1972 Plenum Press, New York

Softcover reprint of the hardcover 1st edition 1972

A Division of Plenum Publishing Corporation

227 West 17th Street, New York, N.Y. 10011

United Kingdom edition published by Plenum Press, London

A Division of Plenum Publishing Company, Ltd.

Davis House (4th Floor), 8 Scrubs Lane, Harlesden, London, NW10 6SE, England

PREFACE

This symposium was held at the 161st ACS National Meeting,
Los Angeles, March/April 1971. It represents a contribution to
the discussion of problems connected with the state of water
near macromolecules. Some papers are only peripheral to the
problem of water structure but may become quite pertinent in
specific cases.

Questions concerned with water structure, rate of hydration,
and similar problems are of importance for biological processes
and are still not yet well understood. It is hoped that the
papers presented here will be of some help in the clarification
of problems in this area.

H. H. G. Jellinek

Department of Chemistry
Clarkson College of Technology
Potsdam, New York
September, 1971

CONTRIBUTORS

S. Ablett, Unilever Research Laboratory, Colworth House,
   Sharnbrook, Bedford, England

M. Anbar, Stanford Research Institute, Menlo Park, California

F. W. Cope, Biochemistry Division, Aerospace Medical Research
   Laboratory, U. S. Naval Air Development Center, Warminster,
   Pennsylvania

B. Crist, Camille Dreyfus Laboratory, Research Triangle Park,
   North Carolina

F. Franks, Unilever Research Laboratory, Colworth House,
   Sharnbrook, Bedford, England

H. R. Gloria, NASA-Ames Research Center, Moffett Field,
   California

G. W. Gross, New Mexico Institute of Mining and Technology,
   Socorro, New Mexico

H. R. Hansen, The Procter & Gamble Company, Miami Valley
   Laboratories, Cincinnati, Ohio

R. S. Kaiser, Research Laboratories, Eastman Kodak Company,
   Rochester, New York

N. Laiken, Department of Chemistry, The University of Oregon,
   Eugene, Oregon

C. E. Lamaze, Camille Dreyfus Laboratory, Research Triangle
   Park, North Carolina

G. N. Ling, Department of Molecular Biology, Pennsylvania
   Hospital, Philadelphia, Pennsylvania

A. P. MacKenzie, Cryobiology Research Institute, Madison,
    Wisconsin

G. Némethy, The Rockefeller University, New York, New York

H. G. Olf, Camille Dreyfus Laboratory, Research Triangle Park,
    North Carolina

A. Peterlin, Camille Dreyfus Laboratory, Research Triangle Park,
    North Carolina

D. H. Rasmussen, Cryobiology Research Institute, Madison,
    Wisconsin

R. L. Reeves, Research Laboratories, Eastman Kodak Company,
    Rochester, New York

R. F. Reinisch, NASA-Ames Research Center, Moffett Field,
    California

G. A. St. John, Stanford Research Institute, Menlo Park,
    California

M. J. Tait, Unilever Research Laboratory, Colworth House,
    Sharnbrook, Bedford, England

H. Yasuda, Camille Dreyfus Laboratory, Research Triangle Park,
    North Carolina

W. Yellin, The Procter & Gamble Company, Miami Valley
    Laboratories, Cincinnati, Ohio

CONTENTS

Introduction

H. H. G. Jellinek

Questions of water structure, water of hydration and similar problems are of biological significance e.g. for the structure of water in living cells, for the role of hydration water around protein molecules in life processes etc. and work on such problems has been done for many years.

The first paper (G. N. Ling) is concerned predominantly with protein hydration. It touches on the controversial problem of "polywater" (anomalous water) as an example of multilayer oriented water. It also refers to the work of T. Hori (1956) who apparently experimented with anomalous water before Derjaguin and his co-workers[1] rediscovered it. Recently, very interesting and careful work by G. Swinzow has been reported[2] dealing with anomalous water. Amongst other things this author has shown that anomalous water cannot be explained away on account of "impurities". It is refreshing to read a report about such careful work after all the hasty experimental work and theoretical speculations which have been published in abundance during the recent past. Ling also points out that hydration of proteins (or macromolecules in general) takes place along the chains at the polymer side groups mentioning work by Dole and Faller of water adsorption on polyvinylpyrrolidone. Work on freezing of aqueous polyvinylpyrrolidone in aqueous solutions[3] is relevant in this connection. Data obtained led to the conclusion that water bound by polyvinylpyrrolidone is "ice-like" having a structure between those of liquid water and ice. The importance of NMR techniques for the problem of water structure is well illustrated in Ling's paper and also plays a very prominent part in most of the other papers of this Symposium. Thus, NMR has a dominant place in the second paper (F. W. Cope) on complexed Na^+ and K^+ ions in biological systems. It is shown that water in cells has definitely more structure than liquid water; however the type of structure cannot be determined by NMR alone and other evidence has to be utilized for solving this problem such as for instance data from adsorption phenomena (polarized multilayers). The next paper (Hansen and Yellin) treats a very specific problem i.e. hydration of stratum corneum. Here, NMR in conjunction with infrared measurements is used. Again there is clear evidence that water is more structured than liquid water in the vicinity of protein molecules.

Hydration of carbohydrate molecules is the topic of the paper by Tait et al. Here NMR is the technique of investigation. Apparently, there are two types of bound water near carbohydrates, one of which is solid-like forming a monolayer on starch (monohydrate). Long range interaction between water and a polysaccharide depends on the type of macromolecule and its concentration in solution.

Peterlin et al. have investigated movement of water in swollen polymers. This work is also of biological significance (e.g. for plants). Here diffusion has been studied as a function of hydration. Diffusion depends on the structure of water and whether it is free, firmly or lightly bound by the macromolecules. NMR was employed in this work and in addition differential calorimetric scanning is used. It is generally found that water in swollen polymers is different from free water; the actual state of this water is difficult to characterize.

Reeves and Kaiser use water soluble dyes as indicators for changes in water structure. These molecules of intermediate size between small and macro-molecules are assumed to have similarities with polymer molecules. This investigation is based on reaction kinetics and spectral measurements. Solvation of a dye molecule will influence its hydrolysis by OH^- ions.

The paper by Laiken and Nemethy has indirect relevance to the main theme of the Symposium but such work may become of importance in specific cases where the presence of water plays a role. The latter influences interaction energies and temperature coefficients for binding. The same is true for the paper by Anbar et al. dealing with hydrated electrons. It is quite feasible that for detailed investigations of hydration of polymers, considerations dealt with in this paper will be of importance.

The paper by Gross is concerned with a different type of problem. It deals with a phenomenon which is now referred to as the Workman-Reynolds effect. This is based on the observation of electric charge separation at an advancing ice/solution interface. Specific ions are preferentially absorbed into growing ice producing freezing potentials which can become of appreciable magnitude. This effect is very likely also operative at a hydration ice layer/solution interface if the solution contains suitable ions, although this is not mentioned by the author. The hydration layer may be that near a protein molecule.

In the paper by Rasmussen and MacKenzie, free energies of ice/solution interfaces as affected by various solutes including polyvinyl-pyrrolidone are discussed. Conclusions can be drawn from changes in these free energies with solute concentration whether the solute is present in excess in the interface. The last paper by MacKenzie and Rasmussen is concerned with sorption and desorption of water on or from this polymer below $0^\circ C$. The paper mentioned above on freezing of aqueous polyvinylpyrrolidone solutions[3] is relevant in this context.

It is hoped that work presented in this Symposium will stimulate further research into the complex problem of water structure, which is so important for life processes.

References

1. B. V. Derjaguin, Scientific Am., _223_, No. 5, 52, (1970).

2. G. K. Swinzow, U. S. Army CRREL, Special Report 156, May, 1971; see also G. K. Swinzow, P. Hoekstra, S. Ackley and W. T. Doyle, Nature, Phys. Science, _229_, 92, (1971).

3. H. H. G. Jellinek and S. Y. Fok, Kolloid Zeitsch. und Zeitsch. fur Polymere, _220_, 122, (1967).

WATER STRUCTURE AT THE WATER–POLYMER INTERFACE

Gilbert N. Ling

Department of Molecular Biology - Pennsylvania Hospital

The fact that multiple layers of gaseous molecules sorb on solid surfaces has long been known. A favorite subject of study has been the condensation of water on glass surfaces (Lehner, 1927; Smith, 1928). In 1929, deBoer and Zwikker (1929) presented a polarization theory, in which they attributed multilayer condensation to a propagated electrical polarization initiated at the polar solid surface. Their theory was presented quantitatively in the form of the following equation:

$$\ln \frac{p_s}{K_3 p_o} = K_2 K_1^{\,a} , \qquad (1)$$

where p_s is the gas pressure, p_o is the saturation gas pressure unaer the same experimental conditions, a is the amount of gas sorbed, and K_1, K_2, and K_3 are constants.

Bradley (1936) derived a similar isotherm, both for gaseous molecules without a permanent dipole moment (e.g., N_2, A) and for those with a permanent dipole moment (e.g., H_2O). Bradley's isotherm is formally identical with the deBoer and Zwikker equation shown above. Bradley showed, for example, that Bray and Draper's data (1926) on the sorption of water on copper oxide fits the equation quite well.

Brunauer, Emmett and Teller, in a well known paper published in 1938, presented the so-called "BET" theory. At the same time, they severely criticized the deBoer-Zwikker-Bradley polarization theory on the grounds that induced polarization in a noble gas like argon is too weak to propagate beyond a monomolecular layer. The BET theory is based on the assumption that polarization by the solid surface is limited to a single layer; additional layers are held together by van der Waal forces operative in normal liquid. The authors were careful to point out, however, that their criticism was limited only to the theory of polarized multilayers of noble gases having no permanent dipole moment and that "if the adsorbed gas has a large permanent dipole moment it is possible that many layers may be successfully polarized by the mechanism of deBoer and Zwikker" (p. 311). Since water has a large dipole moment (1.834 x 10^{-18} e.s.u.) Brunauer, Emmett, and Teller's criticism is clearly inapplicable to the condensation of multilayers of water by induced polarization.

Our next question is how deep the multilayers of water molecules can be. By weighing the water film formed on the surface of titanium dioxide(quartz behaves in a similar manner) and determining the number of water molecules adsorbed per square centimeter, Harkins (1945) calculated the molar energy of desorption minus the energy of vaporization of water in successive layers of water sorbed. He found an excess energy of 6,500 cal/moles for the first layer; 1,380 for the second layer; 220 for the third layer. His data clearly showed that in the case of water sorption on titanium oxide, surface-initiated polarization does not stop at the first layer.

Other experimental data indicate that under certain favorable conditions, an even larger number of water layers can be polarized under the influence of solid surfaces.

THIN WATER LAYERS BETWEEN GLASS SURFACES OR IN NARROW GLASS CAPILLARIES

In 1961, Fedyakins measured the thermal expansion of water in glass capillaries (see Derjaguin, 1970). He found that as long as the inner bores of the capillaries are wider than 1 micron, water expands in the usual manner, displaying a maximum density at 4°C. Below 1 micron, the curves begin to deviate. Indeed, in very fine capillaries, the minimum of specific volume at 4°C disappears altogether, the coefficient of expansion then becoming a constant.

The profound effect of the surrounding surfaces on the water between mica sheets was also shown by N.S. Metsik in connection with the thermal conductivity of water (Derjaguin, 1970). When the water film is wider than 1 micron, the thermal conductivity is as in normal water, below 1 micron, the conductivity sharply rises, so that at a thickness of 0.1 micron the thermal conductivity of the thin water film is more than 10 times higher than that of normal water.

T. Hori (1956), of the Institute of Low Temperature Science of Hokkaido University, measured the freezing point of water between polished glass surfaces and between mica sheets. He found that if the film is 1 mm to 10 microns, the freezing point is -10°C to -30°C. At a thickness below 10 microns, however, there is no indication of freezing even at a temperature as low as -90°C.

Hori also conducted experiments in which he studied the freezing pattern of water films held between a flat glass plate and a curved glass surface with a 35-meter radius of curvature. At a thickness of 2 to 3 microns (as judged by the Newtonian rings), water begins to freeze in an abnormally slow manner. This freezing ceases altogether when the water film is 1μ thick or thinner.

Hori also studied the equilibrium vapor pressure of thin water film held between a flat and a curved glass or quartz surface. The data showed that when the film was more than 1 micron thick, the equilibrium vapor pressure decreased with increasing temperature. Below 1μ the film vapor pressure began to yield a lower vapor pressure at the same temperature than the thicker film. When the films were as thin as 0.1μ there was no detectable vapor pressure even at a temperature as high as 300°C.

Summarizing these important Russian and Japanese findings, we may say that thin films of water between enclosed glass, quartz, or mica surfaces show abnormal behavior when the thickness of these films decreases to 1μ. The departure of the freezing and vaporizing tendencies from those of normal water is profound. Yet, a 0.1μ film is more than 300 water molecules thick!

The radically different properties of a 300-molecule-thick water film between glass surfaces clearly shows that a pair of juxtaposed glass (or quartz) surfaces has a far more profound effect on water than a single surface alone.

A clue to understanding this phenomenon is to be found in the original paper by deBoer and Zwikker (1929). They pointed out that thicker polarized multilayers tend to build up if the solid surface contains alternatingly positive and negative charges. It recently has been pointed out (Ling, 1971) that if such a surface contains both positive (P) and negative sites (N) in a regular array, and, in particular, if two such NP surfaces are brought close together, deep layers of water will exist in an ordered array of dipolar lattices. This follows from the lateral cohesion between the water molecules in the same layer, an interaction shown to equal in magnitude that of the radial cohesion. A stabilized 3-dimensional matrix of polarized water molecules, working in harmony with each other, may then be established. On the other hand, if the surface bears only one type of electrical charge, a lateral repulsion will develop that annuls the radial cohesion. In consequence, little more than one layer of water will adsorb onto unipolar surfaces or sites.

Since it is well known that glass, quartz, and mica surfaces all contain a checkerboard arrangement of N and P sites, the long-range multilayer formations can be so explained.

SOILS AND CLAYS

In recent years, Soviet soil scientists have made rather extensive studies on the water in soils. They have classified the water into three types: suspended water; loosely bound water; and bound water. Suspended water, in its physical properties, is essentially normal liquid water. Bound water is the water in air-dried soils and is no more than 1 or 2 water molecules thick. It is the loosely bound water that, in quantity, is of the greatest importance. Rhode, in a recent monograph, has written: "The remainder of the hull consists of loosely bound water and is formed by a process of multilayer sorption by successive polarization of water dipoles. The thickness of such envelopes may be very considerable - hundreds or even thousands of molecular diameters..." (Rhode, 1969, p. 151). Rhode then points out that "loosely bound water has a diminished capacity for dissolving electrolytes." This property is a matter of considerable importance in relation to the possible role of water structure in the function of living systems (Ling, 1969, 1970).

Evidence in agreement with the polarized multilayer theory of water between clay particles has come from the nuclear magnetic resonance (NMR) studies of Ducros (1960) and of Woessner and Snowden (1968). When clay platelets are oriented in a finite direction, the

water proton NMR signal splits. This splitting is more marked if
one studies the deuteron signal of the D_2O-clay system. Such signal
splitting indicates dipole-dipole interaction of the water or D_2O
molecules that do not average out as in liquid. The NMR signal splitting
is thus in harmony with the concept of deep layers of water between
clay particles, as discussed by the Soviet soil scientists mentioned
above.   It is worthy to note that deuteron signal splitting has been
seen in clays containing as much as 95% D_2O.

PROTEIN HYDRATION

The hydration of proteins has long been a problem of great interest.
A large variety of ingenious hydration methods have been developed
(for review, see Ling, 1971). Yet, in spite of these efforts, there
has also been a persistent doubt whether such hydration exists at
all (Ogston, 1956).

Definitive progress, however, has been made very recently by
Kuntz, Brassfield, Law, and Purcell of Princeton University (1969).
Protein solutes were frozen in liquid nitrogen and the temperature
then raised to -35°C. At this temperature water exists as ice that
yields proton signals too broad to be observed. A certain amount of
water associated with the proteins, however, does not freeze into
normal ice but exhibits a signal much narrower than that of ice
(e.g., 250 Hz), though much broader than that of liquid water (1-2 Hz).
By integrating these water peaks, Kuntz et al obtained the amount of
hydration water, which, most obligingly, agrees well with data from
earlier measurements (0.3-0.5 grams of water per gram of dry proteins).
Thus, the low-temperature NMR studies have finally confirmed the
general concept of hydration and vindicated the employment of the
ingenious indirect methods used to measure its quantities.

Kuntz and his coworkers' NMR findings also clearly and conclusively
proved that the bulk of hydration water is not literally "ice-like,"
for if it were, one should not be able to detect its signal. These
authors further commented that "the lack of a definite freezing point
for the bound water indicates that it is not in a well-organized structure,
even though it is more restricted in mobility than liquid water" (p. 1330).

Another of their noteworthy findings is the unusually large
amount of hydration of denatured t-RNA. At 1.7 g of water per g of
RNA, it is about 4 times greater in quantity than that of the average
protein.

THE SITES OF PROTEIN HYDRATION

Now that we have sound evidence that protein hydration is not merely a figment of the imagination, the question must be raised: what is the mechanism underlying the immobilization of water molecules in the vicinity of protein molecules?

Lloyd and her coworkers (Lloyd and Phillips, 1933) long ago pointed out the interaction of water with proteins through what later became universally referred to as H-bonds, with oxygen, nitrogen, and hydrogen atoms primarily on the protein side chains and to a lesser extent also on the backbone amide groups. Sponsler, Bath, and Ellis (1940) cited a large variety of experimental evidence showing that in a gelatin-gel containing 35% water, about 42% of the water sorbed is on the side chains and the remaining 58% on the backbones.

Further evidence that the backbones offer important sites for anchoring water molecules came from Mellon, Korn and Hoover's (1948) demonstration that polyglycine sorbs water far in excess of that which can be attributed to the amino and ester ends of the polymer and must therefore be due to the backbone amide groups. Dole and Faller (1950) presented similar evidence that synthetic polymers containing no terminal polar groups (e.g., polyvinyl pyrrolidone) nevertheless adsorb considerable amounts of water.

PAULING'S THEORY

Pauling (1945) suggested that only side chain polar groups bind water and that the backbone imino and carboxyl groups do not do so at all. Pauling's theory was based on the BET theory and on the observation that nylon, which in molecular structure resembles the protein backbone, binds very little water.

However, as pointed out by Dole and Faller (1950), sorption of water in polyamides depends not only on the chemical nature of the binding sites but also on the crystallization of the polymer. The failure of nylon to adsorb water can then be ascribed to its high crystallinity.

As pointed out earlier, in the case of sorption of water, which has a high dipole moment, the application of the deBoer-Zwikker-Bradley isotherm seems more appropriate. This view is strengthened by the demonstration of Hoover and Mellon (1957) that water sorption on polyglycine and other polymers can be quantitatively described

by the deBoer-Zwikker-Bradley isotherm up to near saturation,
wheras the BET theory can be fitted to the data satisfactorily only
to about half saturation (Bull, 1944; Ling, 1965).

CONFIRMATION OF THE PAULING THEORY

In spite of the aforementioned criticisms, it would appear that
there has been an increasing amount of experimental evidence supporting Pauling's theory in a general way: (1) Recently Bull and Breeze
(1968) considered the water sorbed at 92% saturation. They suggested
that, with certain qualifications, the data could be described by
assuming that each polar group (excepting side chain amide groups)
binds 6 water molecules and that the backbone binds no water at all;
(2) Fisher (1964, 1965) suggested that each protein molecule in an
aqueous solution tends to assume a spherical shape, with the side
chain polar groups directed outward. The hydration of these surface
polar groups provides a more-or-less uniformly thick envelope of
water molecules around the protein molecules. Fisher then suggested
that there is a stable ratio for the polar outer volume (V_e) and the
non-polar inner volume (V_i). Proteins whose polarity ratio ($p = \dfrac{V_e}{V_i}$)
exceeds this critical value would tend to polymerize with other
molecules to diminish the surface-volume ratio. Proteins whose
polarity ratio falls below the critical value would tend to assume
a non-spherical shape to increase their surface to volume ratio. A
survey of 35 proteins supports this theory.

Since Fisher's theory assumes no backbone hydration, the confirmation of the theory is another indirect support of the original concept
of Pauling.

RESOLVING A PARADOX

Thus, there exists concurrently two diametrically opposed concepts
of protein hydration - one holding that the backbone is the major site
of water sorption, the other holding that it adsorbs no water at all.
And each of these ideas has been steadily gaining support.

It was recently pointed out that a resolution of this paradox becomes
possible if one takes into account the differences in the investigators
and their respective subjects of study (Ling, 1971). The group that
produced evidence in support of the Pauling theory of non-backbone
participation is composed primarily of protein chemists dealing with
aqueous solutions of globular proteins. On the other hand, those

supporting the backbone participation concepts have been primarily industrial chemists involved in the study of fibrous proteins and other synthetic and natural fibers.

Thus, it would seem that for globular proteins the hydration is primarily on surface polar groups to a depth of no greater than 1 layer of water. The fact that these polar groups are primarily monopolar, and therefore do not tend to present an N-P surface, is consistent with this interpretation. The usually large α-helical contents of globular proteins offer de facto interpretations of the lack of hydration of the backbone amide groups.

On the other hand, in fibrous proteins there may be a varying tendency to have part of the backbone existing in the so-called "random coil" state. Under these conditions, the backbone amide groups may adsorb water, to form polarized multilayers, if the geometry of the protein conformation permits.

It is worth noting that of the 13 proteins examined by Bull and Breeze, 9 conform accurately with their theory. All 9 are globular. Of the 3 proteins that do not follow their theory, all hydrate to a much greater degree than theoretically predicted. Of these 3, 2 are fibrous (silk fibroin and collagen). The third, is zein. Mellon, Korn and Hoover (1948) showed that 70% of the sorbed water of this prolamine, zein, was due to the backbone amide groups.

CONCLUDING REMARKS

Water molecules have a large permanent dipole moment as well as a high polarizability. Thus equipped, water tends to interact with polar surface sites at the interface of water and macromolecules. Such an interaction is usually limited to a layer not much more than 1 molecule thick when the surface carries 1 type of electric charge; but, to a depth of hundreds and even thousands of molecules thick when alternating positive and negative sites are presented in a certain geometrical pattern on the surface and when such surfaces are closely juxtaposed. The backbones of proteins can assume either a non-hydrated (e.g., α-helical conformation) or a hydrated (e.g., "random coil" conformation). This flexibility has been regarded as providing the basis of cooperative transitional properties to living protoplasm (Ling, 1969).

SUMMARY

Although hydration of proteins and other macromolecules has long been a subject of interest, much of the earlier evidence came from hydrodynamic studies of protein solutions and the significance of this evidence relies on the validity of assumptions concerning the shape of the protein molecules. Recent NMR studies of frozen protein and nucleic acid solutions have offered unequivocal proof for the reality of hydration and confirmation of earlier conclusions. A survey of the literature reveals that the majority of data in agreement with the Pauling-Bull and Breeze-Fisher theory (a single layer of water molecules are adsorbed on polar side chains only), were derived from the studies of globular proteins. Fibrous proteins, on the other hand, interact with water at the backbone NHCO groups as well. An interpretation for the greater hydration of fibrous proteins was offered in terms of the theory of polarized multilayers. The physiological role of such water existing in polarized multilayers in living cells completes the thesis that hydration of proteins is of prime importance in the molecular biology of living cells.

REFERENCES

1. deBoer, J.H. and C. Zwikker. Z. Physik. Chem. B-3:407, 1929.
2. Bradley, S. J. Chem. Soc. 1467, 1936.
3. Bray, W. C. and H.D. Draper. Proc. Nat. Acad. Sci. 12:295, 1926.
4. Brunauer, S., P.H. Emmett and E. Teller. J. Amer. Chem. Soc. 60:309, 1938.
5. Bull, H. B. J. Amer. Chem. Soc. 66:1499, 1944
6. Bull, H. B. and K. Breeze. Arch. Biochem. Biophys. 128:497, 1968.
7. Derjaguin, B. V. Scientific American. 223:52, 1970.
8. Dole, M. and I. L. Faller. J. Amer. Chem. Soc. 72:414, 1950; see also H.H.G. Jellinek and S. Y. Fok, Kolloid Zeitsch. und Zeitsch. fur Polymere. 220:122, 1967.
9. Ducros, P. Bull. Soc. Franc. Mineral Crist. 83:85, 1960.
10. Fisher, H. F. Proc. Nat. Acad. Sci. 51:1285, 1964.
11. Fisher, H. F. Biochem. Biophys. Acta. 109:544, 1965.
12. Harkins, W. D. Science. 162:292, 1945.
13. Hoover, S. K. and E. F. Mellon. J. Amer. Chem. Soc. 72:2562, 1950.
14. Hori, T. Low Temperature Science A-15:34, 1956. U.S. Army Snow, Ice and Permafrost Establishment, Wilmette, Ill.

15. Kuntz, I.D., Jr., T.S. Brassfield, G.D. Law and G.V. Purcell. Science. 163:1329, 1969.
16. Lehner, S. J. Chem. Soc. p. 272, 1927.
17. Ling, G.N. Ann N.Y. Acad. Sci. 125:401, 1965.
18. Ling, G.N. Intern. Rev. Cytology. 26:1, 1969.
19. Ling, G.N. Intern. J. Neurosci.1:129, 1970.
20. Ling, G.N. "Hydration of Macromolecules" in R.A. Horne's monograph, Equilibrium and Transport Properties of Water and Aqueous Solutions. John Wiley and Sons (1971).
21. Lloyd, D.J. and H. Phillips. Trans. Farad. Soc. 29:132, 1933.
22. Mellon, E.F., A.H. Korn and S.R. Hoover. J. Amer. Chem. Soc. 70:3040, 1948.
23. Ogston, A.G. Proc. Intern. Wool Textile Res. Conf. B:92, 1956.
24. Pauling, L. J. Amer. Chem. Soc. 67:555, 1945.
25. Rhode, A.A. "Theory of Soil Moisture. Vol. 1: Moisture Properties of Soils and Movement". Translated from Russian by the Israel Program for Scientific Translation, Jerusalem, 1969.
26. Smith, J.W. J. Chem. Soc. p. 2045, 1928.
27. Sponsler, O.L., J.D. Bath and J.W. Ellis. J. Phys. Chem. 44:996, 1940.
28. Woessner, D.E. and B.S. Snowden, Jr. J. Chem. Phys. 50:1516, 1969(a).
29. Woessner, D.E. and B.S. Snowden, Jr. J. Coll. Interface Sci. 30:54, 1969 (b).

STRUCTURED WATER AND COMPLEXED Na^+ AND K^+ IN BIOLOGICAL SYSTEMS

Freeman W. Cope

Biochemistry Division
Aerospace Medical Research Laboratory
U. S. Naval Air Development Center
Warminster, Pennsylvania 18974

The living cell is conventionally regarded as a membranous bag containing liquid water with proteins and small ions in free solution. This picture was accepted because of its conceptual simplicity, not because of experimental evidence for its validity. Adequate experimental tests of the basic hypotheses of this picture have become possible only recently.

A minority view of the cell, developed mostly during the last twenty years, and until recently, mostly on the basis of indirect evidence, has held that the cell should be regarded as an organized, non-liquid phase, consisting of a matrix of structured water in which are embedded macromolecules to which are complexed Na^+ and K^+ ions (1-4). The cell may then be considered to resemble a solid, so that cellular ion transport phenomena may be analyzed by the methods of solid state physics (5-7). The observed differences in Na^+ and K^+ concentrations across the cell surface may be understood by regarding the cell as a granule of ion exchange resin, whose complexing sites prefer K^+ to Na^+ and thus maintain a high intracellular concentration of K^+. Water structuring by macromolecules lowers solubility of Na^+ in intracellular water, and thus keeps the concentration of intracellular Na^+ low. Like on an ion exchange resin granule, complexed intracellular ions exchange rapidly with extracellular ions, and energy is not necessary to maintain the difference in ion concentration between granule and surrounding water, so that no ion pumps need be postulated. The last point is of great importance because it avoids the thermodynamic paradox of the conventional picture, which is that the total energy produced by the cell is only a small fraction of that required to maintain the observed gradients of ion concentration against measured rates of ion leakage (1,8).

A clear decision between these two contradictory pictures of cell salt and water metabolism requires definitive experimental evidence on three points, cell water structure, Na^+ complexing, and K^+ complexing. Experimental evidence on the first point is extensive and good, although not absolutely definitive. On the second point it is definitive, and on the third point it is good and promises to be definitive soon because of a new technique.

By nuclear magnetic resonance (NMR), it has been proven that water in animal tissues possesses significantly more structure than liquid water. Steady state NMR studies of H have shown cell water to have a broader NMR spectrum than liquid water, which is consistent with the idea of more structure (9-11). Opponents of the minority view have contended that these broadenings are due to magnetic inhomogeneities in the cell, such as those surmised (but not proven) to be responsible for broadening of the NMR spectrum of water in DNA gels (12). This alternative was already ruled out by Bratton et al (13) who showed by spin echo NMR that T_1 and T_2 of H in muscle water were much shorter than for liquid water, which obviously could not be due to magnetic inhomogeneities, although Bratton et al (13) did not emphasize that point. Short T_2 for water in nerve was shown by Fritz and Swift (14). A possibility remained that these shortenings might be due to paramagnetic ions in cells. The entire question was reinvestigated by Cope (15) using the deuterium nucleus instead of hydrogen to study electric field gradients rather than magnetic dipolar effects in cell water. T_1 and T_2 of D of water in muscle and brain were found to be much shorter than in liquid water, which of course eliminated again the possibility of a magnetic inhomogeneity effect (15). The shortening of T_1 in muscle D_2O was observed to be 200 times too much to be caused by the concentrations of paramagnetic ions present in muscle. The only remaining tenable explanation of short T_1 and T_2 in cell water is high electric field gradients due to increased structure (15).

Although NMR clearly shows that cell water has more structure than liquid water, it does not tell the type of structure. For this information, we must still rely on indirect evidence. I believe the pertinent line of investigation to be that initiated by Ling (16, 17) who pointed out that adsorption of water on proteins conforms to the Bradley adsorption isotherm (18, 19), which implies that polarized water dipoles are adsorbed by proteins in multiple polarized layers, with the degree of ordering decreasing as the radius of the concentric sphere increases. The Bradley isotherm may be written in the form

$$\log[\log(p)] = aV + b \qquad (1)$$

where p is the vapor pressure of water, V is the volume of water absorbed, and a and b are constants. Support for this idea was obtained from the fact that the observed shrinking of cells as a function of solute concentration in surrounding water (the van't Hoff equation) was easily derived from the hypothesis that cell hydration was entirely determined by Bradley adsorption of water on cell proteins (20). The action of solutes in surrounding water was assumed merely to lower extracellular vapor pressure, so that water would desorb from the cell along the Bradley isotherm until equilibrium was reached. In addition to quantitative explanation of experimental data on cell swelling, this approach has the virtue that it does not require any concept of osmotic pressure across the cell surface, a concept which is of dubious validity because of the experimental fact that the cell membrane is freely permeable to solutes which are supposed to exert osmotic pressure across it. Ling (17) has shown that muscle tissue, like purified proteins, also adsorbs water in accord with the Bradley isotherm.

Let us consider the distribution of water structures in the cell. The NMR studies of Cope (15) and of Hazlewood et al (11) have both shown conclusively that total tissue water consists of at least two fractions with different degrees of structuring. These are a small fraction (10-25 percent of total water) with extremely high structuring, and a large fraction (75-90 percent) with a degree of structuring that is (on the average) moderately greater than liquid water. By water adsorption studies on muscle, Ling (17) has confirmed the existence of these two fractions, and has shown that the large fraction adsorbs in accord with the Bradley isotherm, while the small fraction is more tightly bound and has a different adsorption behavior. The NMR behavior of the large fraction is consistent with, although does not prove, the implication of Ling's adsorption study, that the large fraction consists of multiple polarized layers with rapid exchange of protons between layers, thus averaging the different NMR relaxation times of protons in different layers into a single NMR relaxation time for the entire large fraction.

It should be emphasized that the multiple polarized layer structure of cell water is different from ice. It is also different from that of polywater. Polywater, if it exists, is presumed to be stable in pure form. Biological water requires the presence of cellular macromolecules.

Consistent with the concept of structured cell water is recent NMR evidence for cation complexing in the cell. Steady state NMR analysis of Na^+ indicated that 60-70 percent of muscle Na^+ was probably complexed by macromolecules (21, 22). This interpretation was supported by a study using more sophisticated techniques of steady state NMR (23), and was then confirmed by a spin echo NMR study in which the NMR relaxation times of both the free and complexed fractions of muscle Na^+ were measured directly (24). It was shown that 60-70 percent of tissue Na^+ has an extremely short

NMR relaxation time like that of Na$^+$ adsorbed on ion exchange resin, and that the remainder of tissue Na$^+$ has a moderate degree of shortening of NMR relaxation time, as if it were dissolved in structured water (24). NMR has also indicated substantial complexing of Na$^+$ in brain (22, 24), kidney (22, 24), frog skin (25), liver (26), testis (27), and nerve (28).

Preliminary NMR studies suggest that K$^+$ in cells is also complexed (29, 30). When K$^+$ in muscle was exchanged for Na$^+$, the extra Na$^+$ was shown by NMR to be complexed, suggesting the extra Na$^+$ had replaced complexed K$^+$ by competing for common binding sites (29). Direct NMR study of K$^+$ complexing has been difficult due to the low magnetic moment of the ^{39}K nucleus, which requires that a low frequency be used to achieve resonance at the fields obtainable with conventional electromagnets, thus reducing sensitivity below the levels required for the concentrations of K$^+$ present in most biological systems. Nevertheless, a direct NMR approach to cellular K$^+$ complexing has also been made in a preliminary way (30). The problem of instrumental sensitivity was overcome by the use of a high magnetic field (50,000 gauss) from a superconductive magnet and by the analysis of a biological system with an unusually high (1 M) concentration of K$^+$ (the Dead Sea bacterium, H. halobium), the NMR relaxation time of K$^+$ in these bacteria was found to be shortened by at least 50 percent compared with aqueous K$^+$, which is consistent with the concept that K$^+$ is complexed, or is dissolved in structured water, but is not consistent with the idea that bacterial K$^+$ is in free solution in liquid water (30).

Summary

The living cell is pictured as a membranous bag containing liquid water with proteins and small cations in free solution. The cell involved should be regarded as an organized non-liquid phase, consisting of a matrix of structured water with embedded macromolecules complexed with Na$^+$ and K$^+$ ions. Solid state theory may therefore be applied to ion transport phenomena in the cell. Adsorption of water by proteins or muscle conforms to the Bradley isotherm, implying that water dipoles are organized in multiple polarized layers around protein molecules. Recent NMR analysis shows definitely that cell water has more structure than liquid water. The structure of cell water is, however, different from that of ice or of polywater. This view is further supported by strong NMR evidence for complexation of Na$^+$ and K$^+$ ions.

REFERENCES

1. G. N. Ling, Int. Rev. Cytol., <u>26</u>, 1 (1969).
2. G. N. Ling, <u>A Physical Theory of the Living State</u>, Blaisdell, New York (1960).
3. A. S. Troshin, <u>Problems in Cell Permeability</u>, Pergamon Press, London (1966).
4. A. S. Troshin, <u>In Membrane Transport and Metabolism</u> (A. Klein- zeller and A. Kotyk, editors), Academic Press, London (1961).
5. F. W. Cope, Bull. Math. Biophys., <u>27</u>, 99 (1965).
6. F. W. Cope, Bull. Math. Biophys., <u>29</u>, 691 (1967).
7. F. W. Cope, Adv. Biol. Med. Physics, <u>13</u>,1 (1970).
8. G. N. Ling, Amer. J. Phys. Med., <u>34</u>, 89 (1955).
9. E. Odeblad, B. N. Bhar, and G. Lindstrom, Arch. Biochem. Biophys., <u>63</u>, 221 (1956).
10. M. V. Sussman and L. Chin, Science, <u>151</u>, 324 (1967).
11. C. F. Hazlewood, B. L. Nichols, and N. F. Chamberlain, Nature (London), <u>222</u>, 747 (1969).
12. E. A. Balazs, A. A. Bothner-By, and J. Gergely, J. Mol. Biol., <u>1</u>, 147 (1959).
13. C. B. Bratton, A. L. Hopkins, and J. W. Weinberg, Science, <u>147</u>, 738 (1965).
14. O. G. Fritz and T. L. Swift, Biophys. J., <u>7</u>, 675 (1967).
15. F. W. Cope, Biophys, J., <u>9</u>, 303 (1969).
16. G. N. Ling, Ann. N. Y. Acad. Sci., <u>125</u>, 401 (1965).
17. G. N. Ling, Physiol. Chem. and Physics, <u>2</u>, 15 (1970).
18. R. S. Bradley, J. Chem. Soc., page 1467 (1936).
19. R. S. Bradley, J. Chem. Soc., page 1799 (1936).
20. F. W. Cope, Bull. Math. Biophys., <u>29</u>, 583 (1967).
21. F. W. Cope, Proc. Nat. Acad. Sci. (USA), <u>54</u>, 225 (1965).
22. F. W. Cope, J. Gen. Physiol., <u>50</u>, 1353 (1967).
23. J. L. Czeisler, O. G. Fritz, and T. L. Swift, Biophys. J., <u>10</u>, 260 (1970).
24. F. W. Cope, Biophys. J., <u>10</u>, 843 (1970).
25. C. A. Rotunno, V. Kowalewski, and M. Cereijido, Biochim. Biophys. Acta., <u>135</u>, 170 (1967).
26. D. Martinez, A. A. Silvida, and R. M. Stokes, Biophys. J., <u>9</u>, 1256 (1969).
27. I. L. Reisin, C. A. Rotunno, L. Corch, V. Kowalewski, and M. Cereijido, Physiol. Chem. and Physics, <u>2</u>, 171 (1970).
28. F. W. Cope, Physiol. Chem. and Physics, <u>2</u>, 545 (1970).
29. G. N. Ling and F. W. Cope, Science, <u>163</u>, 1335 (1969).
30. F. W. Cope and R. Damadian, Nature (London), <u>228</u>, 76 (1970).

NMR AND INFRARED SPECTROSCOPIC STUDIES OF STRATUM CORNEUM

HYDRATION

J. R. Hansen and W. Yellin

The Procter & Gamble Company, Miami Valley Laboratories,

Cincinnati, Ohio 45239

INTRODUCTION

Nuclear magnetic resonance (nmr) and infrared (ir) spectroscopy have been used to study the state of water in a variety of
biological systems.[1-8] In our study of the state of water in
in vitro human stratum corneum (s.c.), the outer layer of the
epidermis, we have combined these two techniques. Ir and nmr
spectroscopy are complementary in the sense that pulsed nmr can
provide a direct measure of molecular mobility,[1,9] while ir
spectroscopy can provide information on the strength of intermolecular hydrogen-bonding.[8] Thus, both motion and order, which
may be largely independent, can be compared in the same system.

EXPERIMENTAL

The first figure is a schematic representation of the human
skin in cross-section. The top layer is the s.c., about 10 μ in
thickness, representing about 1% of the total skin thickness. It
is composed of flattened, keratinized cells, each of which is about
0.5 μ thick and about 40 μ in diameter. Dry s.c. contains about
75% protein, 25% lipid, and small amounts of low molecular weight
substances. The s.c. cell is enclosed in a chemically-resistant
membrane and contains a mixture of fibrous proteins, amorphous
proteins, and small molecular weight substances.

The s.c. samples were obtained from human hip skin by means
of dermatome. Viable cells were removed from the excised skin by
trypsin digestion. The resultant s.c. pieces were stored under
vacuum over desiccant until used. The sample preparation method

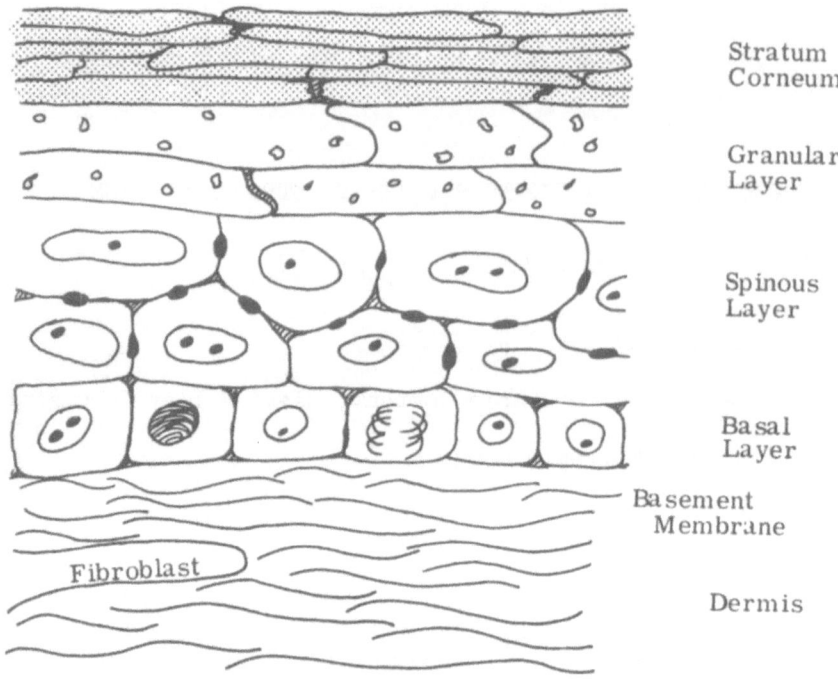

Figure 1. Schematic representation of the human epidermis in
           cross-section.

was designed to ensure that the samples were as close as possible
to the <u>in vivo</u> state.

The sample preparation method employed was to hydrate for
various time intervals the dermatome isolated samples of stratum
corneum. For the ir measurements, hydration was effected via the
vapor phase from a 5/95%, D_2O/H_2O, salt solution (96% relative
humidity to prevent condensation and adsorption of liquid water).
Pure H_2O vapor hydration was used for the nmr experiments. Ir
spectra were obtained on a Perkin-Elmer 421 grating spectrometer,
equipped with a Beckman LTV-2 low temperature cell. Nmr relaxation
times were measured on a Magnion ELH-30 pulsed spectrometer at
36 MHz, using the standard Carr-Purcell techniques.[10]

RESULTS AND DISCUSSION

The ir spectrum of H_2O is complicated by the fact that the OH
stretch is coupled to other modes of motion and is therefore
difficult to interpret. We therefore hydrated the s.c. samples
with a 5% D_2O/95% H_2O solution, and looked at the uncoupled OD

oscillator (HOD surrounded by H_2O molecules), which is not complicated by coupling to other modes of motion. We have previously found[11] that for bulk water the absorption band of the uncoupled OD oscillator is a broad Gaussian band in the liquid phase and a narrower Lorentzian band in ice. Figure 2 shows a schematic of the ir spectrum of HOD in s.c. At +30°C, the uncoupled OD oscillator appears as a broad asymmetric band, centered at 2502 cm^{-1}, similar to the OD band in the bulk liquid, except it is skewed towards lower frequencies. To further simplify the spectrum, the sample temperature was dropped to -50°C, at which point a sharp Lorentzian band appears, similar to that of ice, and which we attribute to water which is not structured by interaction with the protein and can therefore freeze. The appearance of this ice band (2432 cm^{-1}) at -50°C occurs with a corresponding loss in intensity of the broad band at 2502 cm^{-1}. By observing the OD band at low temperature, we can then separate the ir absorption into 2 components: water which freezes at -50°C (narrow Lorentzian component), and water which is constrained, presumably by

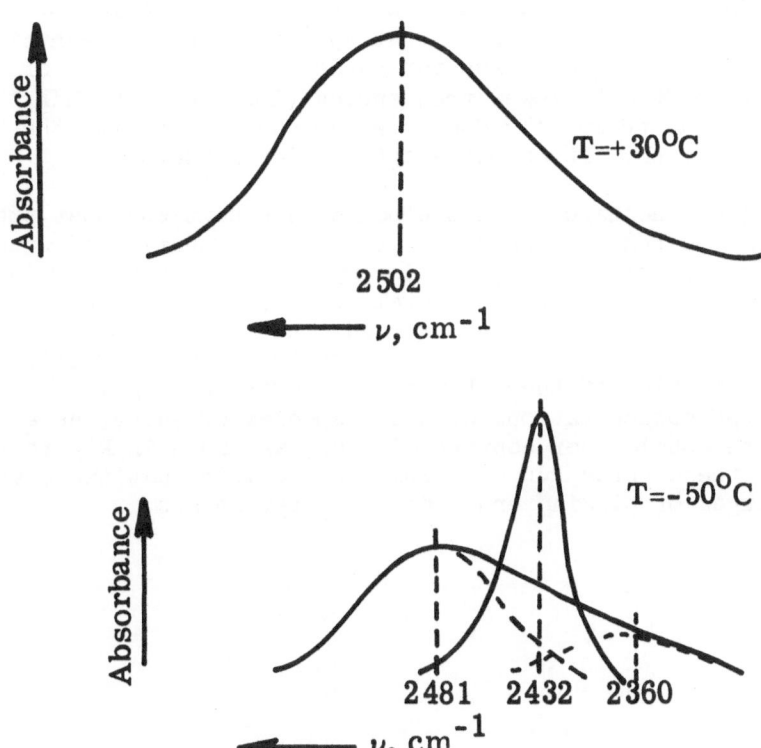

Figure 2. Schematic ir spectra of OD oscillators for HOD in
s.c. at 30°C and -50°C.

interaction with the s.c. protein, such that it cannot assume the ice structure (broad, asymmetric Gaussian band).

The exact temperature at which this Lorentzian component appears seems to depend on the degree of hydration of the sample and on the rate of cooling, but it usually appears at -30°C and has reached its maximum amplitude at -50°C. Figure 3 shows ir spectra of the OD region at 30%C and as a function of time after cooling to -50°C. The sharp Lorentzian band due to ice is not present at +30°C, and grows in magnitude with time after cooling to -50°C.

When the ice-like Lorentzian band is subtracted from the spectrum (using a DuPont 310 curve resolver), a residual broad absorption remains, extending from 2500-2300 cm^{-1}. This absorption persists even at liquid nitrogen temperature, but it can be removed by desiccating the sample. Thus the absorption is due to a species of HOD and not to ir absorption of ND (that is, not to replacement of amide protons by deuterium). This residual broad, asymmetric absorption can be consistently decomposed into 2 broad, Gaussian bands at 2481 and 2360 cm^{-1}, which are similar in shape to liquid HOD bands. Three bands in the OD region of HOD in s.c. are then distinguishable at -50°C: a sharp Lorentzian band due to ice, and 2 broad Gaussian bands similar to those of liquid HOD but shifted to lower frequencies (implying a higher energy H-bond), and presumably due to water associated with the s.c. in such a way that it cannot adopt the ice structure.

Using these lineshapes plus the boundary conditions that the peak positions and band widths be invariant as a function of water content, it is possible to uniquely decompose the ir spectrum into 3 bands over a range of hydration levels by varying only the relative intensities. The bands observed, spectral parameters, and assignments in terms of water are given in Table 1. The ir absorption intensities of these bands are converted to give estimates of the concentrations of the 3 species of water, as a function of the total water content of s.c., as shown in Figure 4. We think that these three HOD absorption bands mean that there are 3 types or ranges of binding energies for water in s.c.

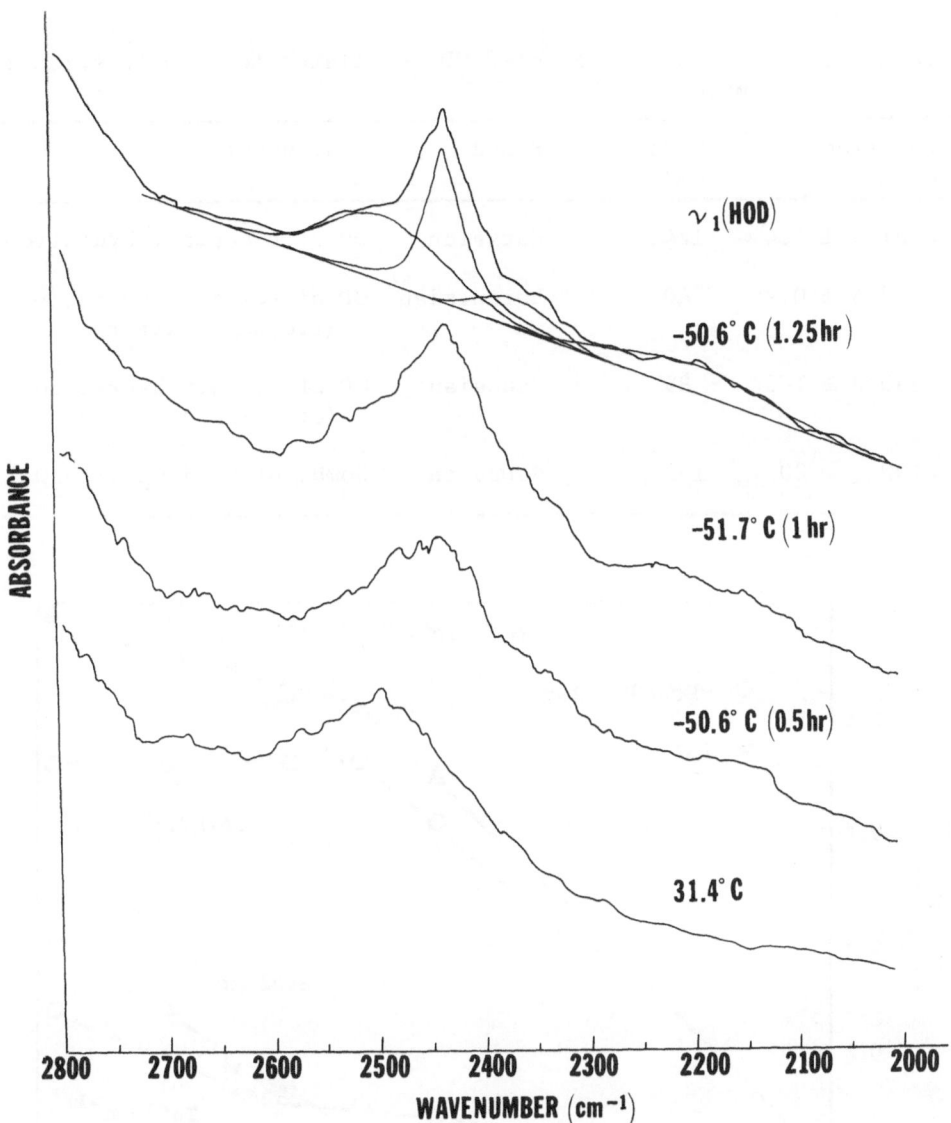

Figure 3. Ir spectra of OD oscillators at 30°C and as a function of time after cooling to -50°C, for a s.c. sample containing 59% water.

Table 1.   IR bands of uncoupled OD oscillator for HOD in stratum
           corneum

Position cm^{-1}	Width cm^{-1}	Shape	Assignment
2481.4 ± 1.4	124	Gaussian	OD of secondary hydration
2432.4 ± 0.9	40	Lorentzian	OD of tertiary hydration (freezable water)
2360.0 ± 1.6	83	Gaussian	OD of primary hydration (at polar sites)
2190 ± 20	163	Gaussian	Comb. of $\nu_2 + \nu_R$ of H_2O

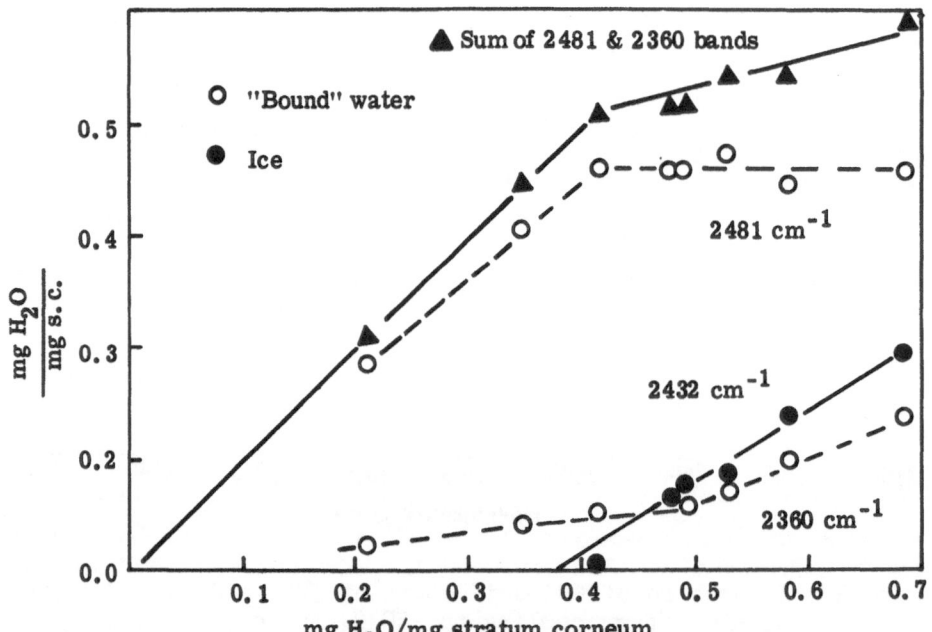

Figure 4.   Water species distribution in s.c., showing weight
            fraction of water of each type present in s.c. as a
            function of total water content.

The three types of water in stratum corneum are:

(i)   A tightly bound HOD absorbed at polar sites of the
       protein, e.g., carboxyl or amine groups (the broad
       Gaussian band at lowest frequency, 2360 cm^{-1}, highest
       binding energy), present at all hydration levels. We
       call this primary water of hydration.

(ii)   A less tightly bound HOD, probably representing water
       hydrogen-bonded to the primary water of hydration, or
       water bound to dipolar sites on the protein (the
       second broad Gaussian band at higher frequency,
       2481 cm^{-1}) increasing in amount up to about 40% total
       water content and then appearing to level off in
       concentration.

(iii) A third species of water whose interactions with the
       s.c. protein are so weak that it can take on the ice
       structure, e.g. freeze, between -30 and -50°C. This
       species is the most like bulk liquid water of the 3
       species. It does not appear until ~ 40% total water
       content is reached (the same concentration at which
       the secondary water of hydration sites are saturated),
       and then increases linearly with total water content.

Thus, the primary and secondary binding sites must be filled before
the freezable water appears. The primary and secondary waters of
hydration account for the initial $\approx$ 40% of water taken up by the
s.c. There is a suggestion of change in protein conformation upon
hydration. The increase in the population of primary water of
hydration above 50% total water content may mean an unfolding of
the protein to expose additional high energy binding sites.

We have seen that by ir measurements it is possible to divide
the water present in s.c. into two main types: water which is
associated with the s.c. protein such that it cannot assume the
ice structure at -50°C; and water which interacts so weakly with
the protein that it can freeze between -30 and -50°C. These same
effects can also be distinguished from measurement of the water
molecular mobility by pulsed nmr. The basic theoretical interpretation of the nmr relaxation time results is fairly simple. We
have measured the rate at which the proton transverse magnetization
of the water in s.c. returns to equilibrium after an initial
perturbation. This recovery can be described by a relaxation
time (or relaxation times). In the temperature range under study,
the longer the relaxation time, T_2, the greater the water molecular
mobility.

Figure 5 shows the spin-spin relaxation curve for water in a
s.c. sample containing about 100% weight fraction water at 30°C.

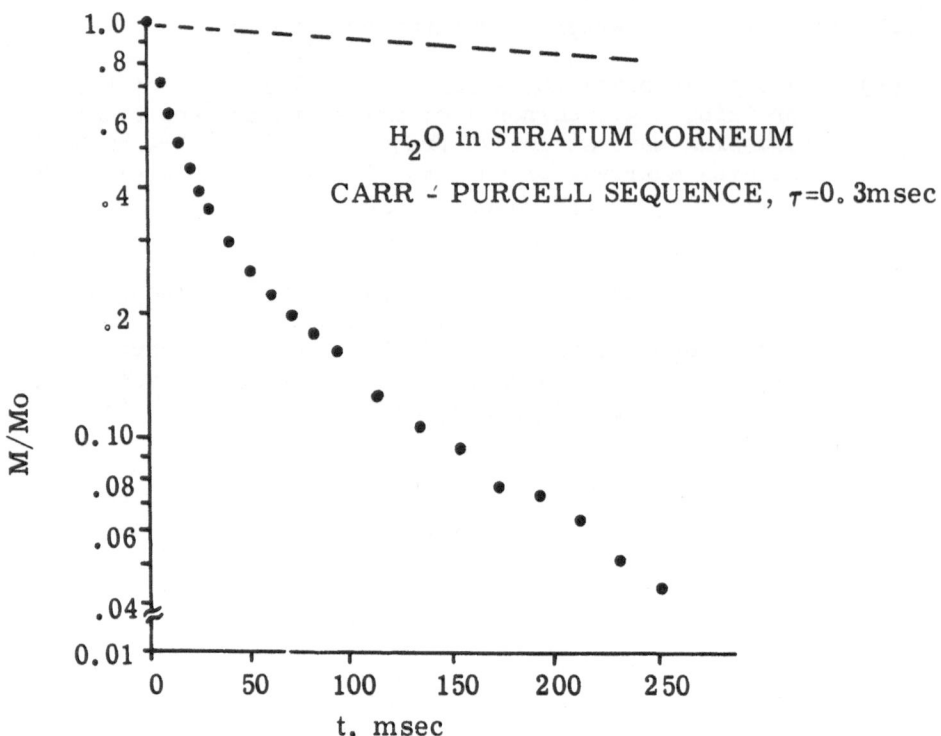

Figure 5.   Spin-spin relaxation of water in s.c. at 30°C:
            Carr-Purcell sequence.

The dashed line at the top shows the same curve for bulk liquid
water. There are two points to be noted from this data. The
s.c. water relaxation curve is not describable by a single
exponential term, as is the bulk liquid water. This means that
there exist more than one type of water in s.c., with differing
molecular mobilities, and exchange between the two types is slow
on the nmr time scale.[12] In fact, the relaxation data for water
in s.c. can be interpreted by assuming the existence of two
species of water, one with a relaxation time T_2 of 17 msec and
the other with a T_2 of 106 msec. These relaxation data for
samples containing 50, 100, and 150% weight fractions of water
can be fit assuming two species of water, with spin-spin relaxation
times 17 and 106 msec. The water species with 17 msec relaxation
time (the less mobile species) is present to the extent of 30 ± 5%,
based on the dry s.c. weight, while the more mobile fraction, with
T_2 of 106 msec, increases with water content, above 30% total water.
In addition to the existence of two species of water, a second
piece of information comes from the actual values of the observed
relaxation time. The relaxation time for bulk liquid water at
30°C is about 3 sec, and for ice is about 30 μsec. Thus the
relaxation time of 17 msec observed for the less mobile fraction

of water in s.c. is intermediate between the liquid and ice values, while the T_2 of 106 msec for the more mobile fraction is $\approx$ 6X longer. Both water species have relaxation times less than the bulk liquid. Whether this is due to restriction of the translational or the rotational diffusion cannot be determined from these experiments. If the relaxation of water at -50°C is observed, the more mobile (longer T_2) species of water disappears due to freezing (ice is not observable with our apparatus). The nmr and ir techniques then give the same result in terms of freezing of the more liquid-like water species.

A comparison of the ir and nmr data on water in s.c. shows that the first 30-40% of water sorbed strongly interacts with the s.c., as evidenced both by nmr and ir measurements, while the water present above this amount is more like bulk liquid water in its properties. There are two possible reasons for the observation of two "bound" water species by ir and only one by nmr. The primary water of hydration as observed by ir (the band centered at 2360 cm^{-1}) may have a relaxation time so short (molecular reorientation rate so reduced) as to be unobservable by our nmr apparatus (i.e., $T_2 \gtrsim 40$ μs).[1] The second possibility is that water exchange is rapid between the primary and secondary hydration sites. This exchange could be rapid on the nmr time scale (10^{-8} sec), resulting in averaging of the nmr parameters,[13] but slow on the ir time scale (about 10^{-14} sec), resulting in distinct resolution. In order to determine if a non-exchanging, highly immobilized water species exists that is not observable with our pulsed nmr apparatus, wide line nmr experiments were performed on D_2O in s.c. In this case a broad line due to the tightly bound water, with a narrow line due to secondary water of hydration superimposed on it would be expected. Sweeps of up to 200 gauss in width revealed only a single D resonance line for a s.c. sample containing 50% water. Thus we conclude that exchange of the primary and secondary water of hydration as observed by ir results in averaging of the nmr relaxation behavior.

Summary

IR and nmr evidence for the existence of three species of water in hydrated human stratum corneum, with differing intermolecular hydrogen-bonding strengths and molecular mobilities, have been presented. The two "bound" water fractions observed by ir apparently exchange rapidly on the nmr time scale and are not distinguishable by nmr. The third, bulk liquid-like water species is distinguishable from the bound water by both ir and nmr. The water associated with the s.c. was found to be present to the extent of 30-40%, based on the dry s.c. weight.

REFERENCES

1. Cope, F. W., Biophys. J., 9, 303 (1969).

2. Dehl, R. C. and Haine, C. A. F., J. Chem. Phys., 50, 3245 (1969).

3. Kuntz, I. D., Brassfield, T. S., Law, G. D., and Purcell, G. V., Science, 163, 1329 (1969).

4. Glasel, J. A., Nature, 218, 953 (1968).

5. Kruger, G. J., Helcke, G. A., Magnetic Resonance and Relaxation (edit. by Blinc, R., Hadzi, D., and Osredkar, M.), 1136 (North-Holland, Amsterdam, 1967).

6. Hazelwood, C. F., Nichols, B. L., and Chamberlain, N. F., Nature, 222, 747 (1969).

7. Woessner, D. E. and Snowden, B. S., Jr., J. Coll. Interface Sci., 34, 290 (1970).

8. Falk, M., Poole, A. G., and Seymour, C. G., Can. J. Chem., 48, 1536 (1970).

9. Hansen, J. R. and Lawson, K. D., Nature, 225, 542 (1970).

10. Carr, H. Y. and Purcell, E. M., Phys. Rev., 94, 630 (1954).

11. Yellin, W. and Courchene, W. L., Nature, 219, 852 (1968).

12. Zimmerman, J. R. and Lasater, J. A., J. Phys. Chem., 62, 1157 (1958).

13. Allerhand, A. and Gutowsky, H. S., J. Chem. Phys., 41, 2115 (1964).

AN NMR INVESTIGATION OF WATER IN CARBOHYDRATE SYSTEMS

M. J. Tait, S. Ablett, and F. Franks

Unilever Research Laboratory, Colworth House

Sharnbrook, Bedford, England

The importance of protein-water interactions and the suitability of the Nuclear Magnetic Resonance technique for characterising the state of the aqueous component has already been discussed at this Symposium. We shall report a complementary study of the state of water in carbohydrate systems and indicate to what extent water plays a significant role in these systems.

WATER ON STARCH

It is useful to first establish the state of water adjacent to a carbohydrate surface in order to discover whether its properties differ significantly from those of bulk water, and for this purpose, a low moisture content starch was investigated. The deuteron Magnetic Resonance spectra of samples hydrated with D_2O were recorded in preference to the more usual Proton Magnetic Resonance studies for two reasons:-

(i)   In low moisture content systems, the DMR spectra are likely to be less sensitive to "proton" or water molecule exchange

(ii) Deuteron relaxation is not influenced by restricted diffusion of water molecules (1).

Since both of these effects complicate the interpretation of the spectra in terms of the state of the water, they are best avoided if possible.

The observed DMR spectrum for samples of moisture contents up to 20% was a single line. For a sample containing 20% moisture, the integrated intensity of the signal is shown as a function of temperature in Fig.1. It can be seen that there is

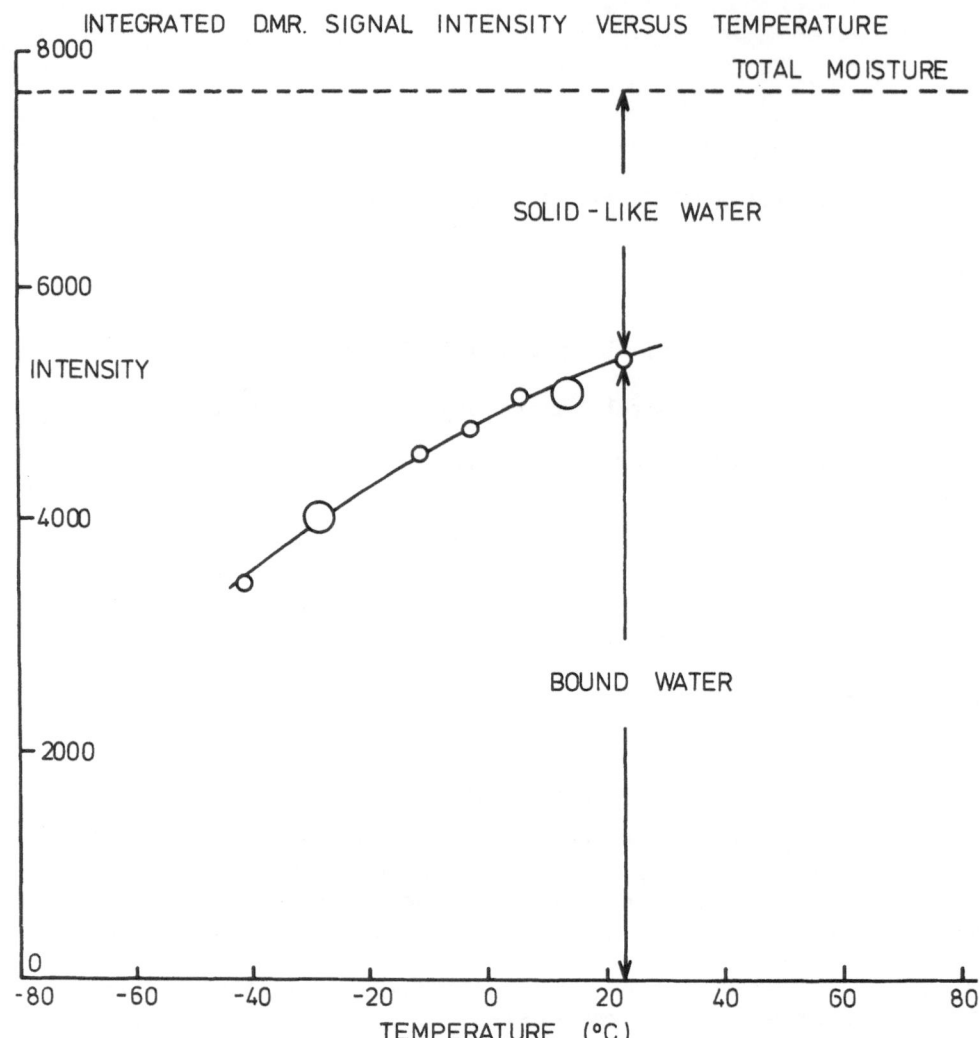

Fig. 1. D_2O adsorbed on starch.

no sharp decrease in the intensity at the freezing point, nor
at any temperature down to $-40°C$, i.e. there is no free water
present (2), furthermore some of the water present in the sample
does not contribute to the signal.   It follows that the
observed signal is due totally to bound water, and some water
is present in an even more firmly bonded condition, such that it
gives rise to a very broad and consequently unobserved DMR
signal.   This fraction is referred to as solid-like water in Fig.1.

The concentration of solid-like water, evaluated as in Fig.1, was independent of moisture content over the range 6-20%. At moisture contents below about 4%, no DMR signal was detected in a single scan, but the solid-like water signal of intensity corresponding to the analytical moisture content did appear on improving the signal-to-noise ratio by repetitively sweeping the spectral region and accumulating the signal with a CAT. So, in effect, two states of water can be resolved.

The solid-like water concentration was found to be in good agreement with the monolayer concentration of $7 \pm 1\%$, calculated by applying the B.E.T. treatment (3) to the sorption isotherm. Moreover, the DMR results show that this monolayer concentration shows no significant increase for moisture contents of up to three times this monolayer value, which suggests that the monolayer or solid-like water completely occupies the specific hydration sites in the starch. The molar ratio of water to starch monomer at the monolayer concentration is 0.7, corresponding approximately to a monohydrate, and it has been concluded (4) that, statistically, two of the three OH groups per monomer of starch are not available for hydrogen bonding to water. A recent X-ray study (5) of hydrated starch (B amylose) suggests that the $O(2)H$ and $O(3)H$ hydroxyl groups hydrogen bond to contiguous residues, and the remaining $O(6)H$ group forms an interchain H-bond involving a water molecule, i.e. $O(6)H-H_2O-O(2)H$.

The linewidth of the bound water signal is shown as a function of temperature in Fig.2 for the sample containing 20% moisture. The linewidths were independent of a two-fold change of magnetic field intensity and therefore are not determined by local inhomogeneities in the sample. It was also confirmed that at $20^{\circ}C$ T_1 was certainly very short and of the same order of magnitude as the T_2 calculated from the linewidth. The linewidth was much greater than that of bulk water — as should be the case for a bound water signal — but there is a sharp change in the linewidth at about $4^{\circ}C$. This indicates that the water-water interactions must be sufficiently extensive to enable them to undergo a co-operative breakdown at the melting point of D_2O. Apparently the mobility of the bound water molecules is restricted due to hydrogen bonding with other water molecules rather than with the starch. From the linewidths, it can be calculated (6) that the rotational frequency of the molecules in this bound water fraction is reduced by a factor of 10^4-10^5 compared to that in pure water. Consequently the properties of water molecules even once removed from contact with the polysaccharide are considerably modified.

WATER IN POLYSACCHARIDE GELS

NMR studies on water in polysaccharide gels, which can contain up to 99% water, were, in the early days (7,8) also

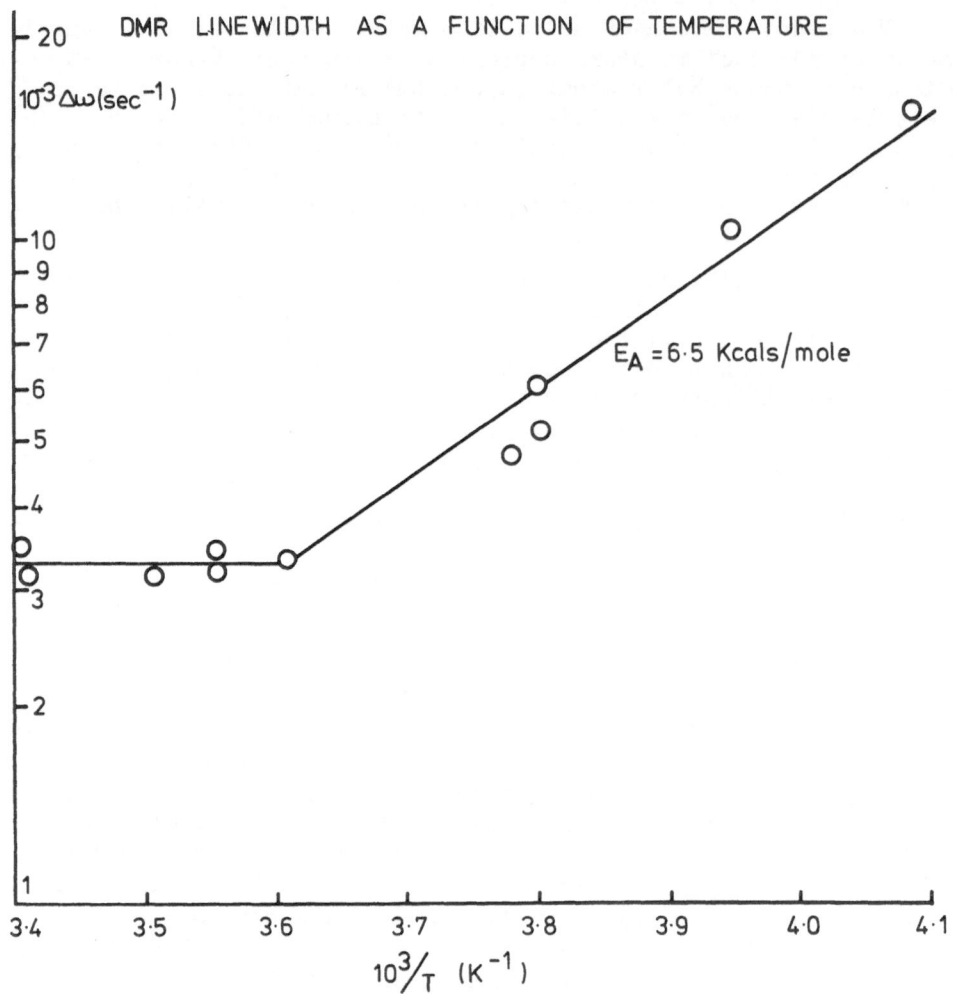

Fig. 2. D_2O adsorbed on starch.

thought to indicate a highly restricted mobility of the water molecules.   However, these studies were based on proton linewidth measurements which, as pointed out previously, can be misleading.   We have repeated these measurements (9) on Agarose, which sets to a gel on cooling and has the disaccharide repeating unit shown in Fig.3.   It was immediately clear from these studies that if proton T_1 measurements were employed rather than T_2 as a criterion for water structuring, then a very much smaller degree of ordering of the water molecules is apparent.   Some preliminary results, obtained recently, which confirm this interpretation are given in Fig.4. These measurements refer to T_1 relaxation of ^{17}O

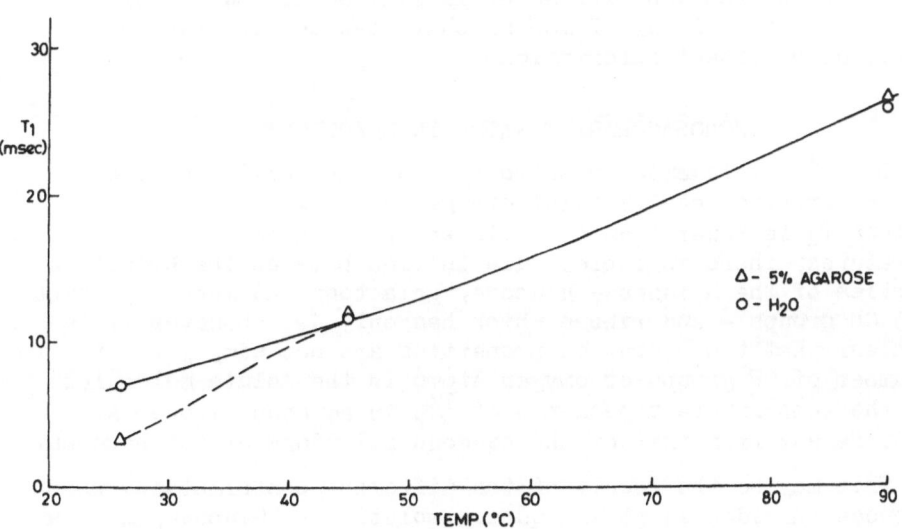

Fig. 3.   Repeating unit of agarose.

Fig. 4. T_1 for $H_2{}^{17}O$ in 5% agarose.

in enriched water and have advantages over both proton and
deuteron studies by being completely independent of "proton"
exchange effects and, a factor of at least 100 times more
sensitive to changes in rotational motion of the water molecules.
It can be seen from Fig.4 that at temperatures above the gel
point, 43°C, there is no detectable difference between the
rotational properties of water in Agarose solution and those of
bulk water.   In the gel state, T_1 is lower than for bulk water
but it has been shown recently (10), that even in the gel, the
self diffusion coefficient for water is only reduced by 15% and
that the best estimate of water bound to the polysaccharide (11)
is only one molecule per repeating unit.

Formally there is thus some resemblance to starch in so far
as the monohydrate appears to be stable in the gel state but,
unlike starch, there is no detectable long range interaction
between water and the polysaccharide at this concentration in
a gel.   However, the Agarose solution is fairly dilute. Its
repeating unit concentration is only 0.02 molal, and it is well
known that in solutions the identity of hydrated species is
blurred by the frequent exchange of the water molecules in the
hydration shell with those in the bulk solvent.   To weight the
time-averaged result more in favour of the hydrated species,
it is necessary to raise the concentration of solute.   The
polysaccharides themselves have a relatively low solubility, but
since their interaction with water is limited to individual
repeating units, a study of monosaccharide-water interactions
can provide pertinent information.

MONOSACCHARIDE-WATER INTERACTIONS

Some ^{17}O relaxation results for monosaccharide solutions
at a concentration of 2.7 molal are given in Fig.5.   In all
solutions T_2 is lower than it is in water.   At lower
temperatures, there is a clear distinction between the hydration
properties of the hexoses – glucose, galactose and mannose, which
have 5 OH groups – and ribose which has only 4.   However it is
also clear that the hydration properties are not simply related to
the number of OH groups or oxygen atoms in the solute molecules
since the temperature dependence of $1/T_2$ in aqueous ribose is
very different from that of the aqueous solutions of the hexoses.

This may be the result of the different conformations which
the monosaccharides adopt in aqueous solution.   Glucose, mannose
and galactose have a C1 conformation (12).   The majority of the
OH groups are in an equatorial configuration and the distance
between the groups, in this configuration, or the same side of
the ring, e.g. OH groups attached to carbon atoms 1, 3 and 5 of
β-D glucose, is 4.86Å (13).   This is compatible with the most
probable (14) second nearest neighbour distance in water of 4.9Å,
which is derived from the X-ray data for water assuming a

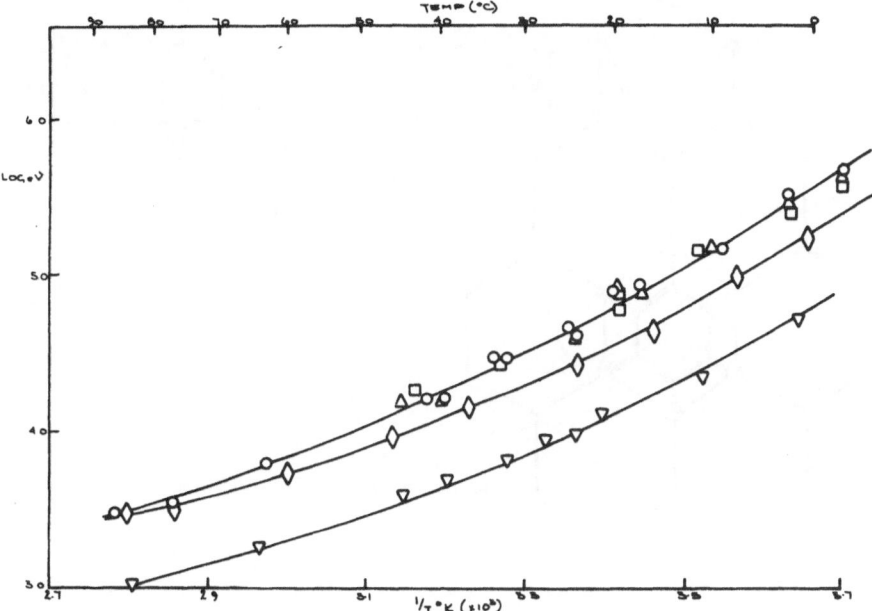

Fig. 5. Linewidth of $H_2{}^{17}O$ in monosaccharide solutions,
corrected for inhomogeneity as a function of
temperature. ∇ ^{17}O-enriched water; ◇ 2.69 M
ribose; △ 2.79 M galactose; ○ 2.79 M glucose;
□ 2.79 M mannose.

tetrahedral structure, and supports the proposal (15) that
hydrogen bonding would be expected between water and the majority
of the hexose OH groups.   For example, each of the OH groups
of β–D glucose is in an equatorial configuration and fits very
well with the tetrahedral arrangement of water molecules (Fig.6).
Ribose however, (12) adopts 1C and furanose conformations in
aqueous solution in addition to C1.   Less than half of the OH
groups of the 1C and furanose configurations can hydrogen bond to
water by means of replacing water molecules as proposed in Fig.6.
Estimating hydration according to how well the monosaccharide
OH groups can be accommodated by a three dimensional version of
the model in Fig.6, gives an average ratio of 2:1 for hexose:
ribose.

As the temperature is raised, the radial distribution function
for water, obtained from the X-ray diffraction pattern (16), shows
that there is a decreasing probability of water molecules being
spaced 4.5–5.3Å apart;   the peak in the radial distribution
function corresponding to this separation has virtually

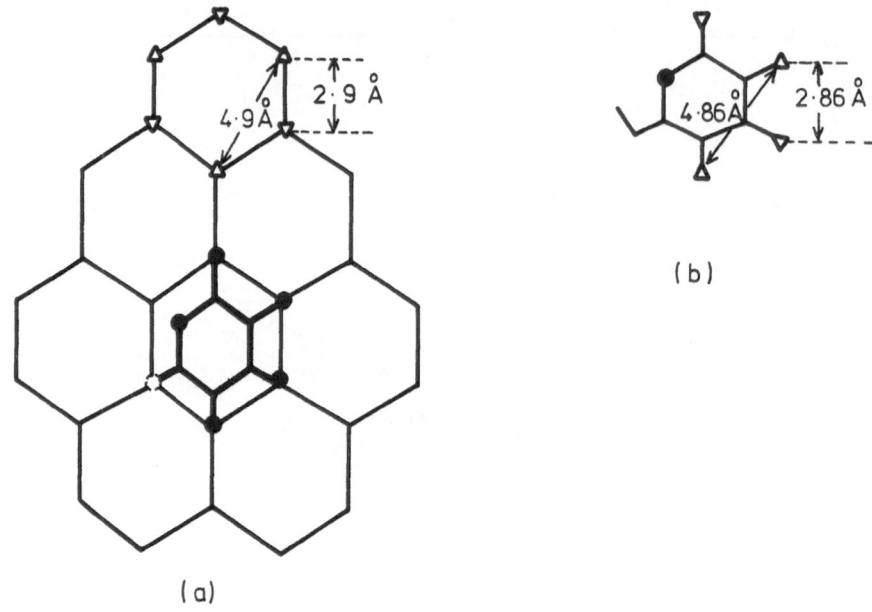

(b)

(a)

Fig.6. Possible model for the hydration of monosaccharides.
(a) Tridymite water structure at 25°C, (b) β–D Glucose
The orientation of the triangles indicate whether an
oxygen atom is above or below the plane of the ring.

disappeared at 75°C.   The difference between the hydration
properties of the hexoses and ribose also fades at higher
temperature (Fig.5) and at 80°C there is no clear distribution
between the two types of solute.   However, even at 80°C, $1/T_2$ of
water in these solutions is not the same as in the pure liquid,
and a possible explanation is that these monosaccharides form
stable monohydrates in aqueous solution.   Glucose and galactose
monohydrates have been isolated and the melting points of the
monohydration are 83°C and 118–120°C respectively (17).

It is known that a higher proportion of the furanose
conformations are found in non aqueous solvents, and it has been
suggested (18,12) that the 5 membered ring conformation is

inherently the more stable in the absence of solvation.   It follows that the pyranose forms are stabilised by interaction with water, and our NMR measurements provide evidence for the greater hydration of the 6-membered ring conformation and offer a description of the nature of the hydrophilic interaction.

SUMMARY

1. Particular conformations of monosaccharides are stabilised by what is thought to be the structure of bulk water.

2. With polysaccharides this facility is lost because the conformation of monomer units is fixed in the polymer. The gross reduction in rotational freedom of the water molecules in Agarose gel, which had been inferred from earlier NMR studies, is shown not to occur, and, in these fairly dilute systems, the only bound water which NMR can detect corresponds to one water molecule per residue.

3. A monohydrate is also indicated when water is adsorbed onto starch, but, in this case, the properties of water molecules not directly bonded to the starch are also considerably modified.   This factor will be important in describing the gelatinisation of starch, a term which refers to the swelling of starch granules, but apparently does not contribute to the stability of polysaccharide gels.

REFERENCES

1.   J.A. Glasel, Nature, 227,   705 (1970).

2.   R. Toledo, M.P. Steinberg & A.I. Nelson, J. Food Sci., 33, 315 (1968).

3.   S. Brunauer, P.H. Emmett and E. Teller, J. Am.Chem. Soc, 60, 309 (1938).

4.   M. Masuzawa and C. Sterling, J. Appl. Polymer Sci, 12, 2023 (1968).

5.   J. Blackwell, A. Sarko and R.H. Marchessault, J. Mol. Biol., 42, 379 (1969).

6.   H.G. Hertz, "Progress in NMR Spectroscopy," ed. Elmsley, Feeney and Sutcliffe, Pergamon Press, 3, (1967).

7.   D. Hechter, T. Wittstruck, N. McNiven and G. Lester, Proc. Nat. Acad. Sci. U.S.A., 46, 783 (1960).

8.   C. Sterling and M. Masuzawa, Makromol. Chem., 116,   140 (1968)

9.   T.F. Child, N.G. Pryce, M.J. Tait and S. Ablett, Chem. Comm., 1214 (1970).

10.   D.E. Woessner, B.S. Snowdon and Y.C. Chiu, J. Coll. Interface Sci., 34, 283 (1970).

11. D.E. Woessner, B.S. Snowdon, J.Coll. Interface Sci., 34, 290 (1970).

12. S.J. Angyal, Angewandte Chemie, Int. Edit., 8, 157 (1969).

13. D.T. Warner, Nature, 196, 1055 (1962).

14. M.D. Danford and H.A. Levy, J. Am. Chem. Soc., 84, 3965 (1962)

15. M.A. Kabayama, D. Patterson and L. Piche, Can. J. Chem., 36, 557, 563 (1958).

16. A.H. Narton, M.D. Danford and H.A. Levy, Disc. Faraday Soc., 43, 97 (1967).

17. Dictionary of Organic Compounds, Eyre and Spottiswoode, 3 (1965).

18. R.V. Lemieux and J.D. Stevens, Can. J. Chem., 44, 249 (1966).

MOVEMENT OF WATER IN HOMOGENEOUS WATER-SWOLLEN POLYMERS

H. Yasuda, H. G. Olf, B. Crist, C. E. Lamaze,
and A. Peterlin

Camille Dreyfus Laboratory, Research Triangle Institute

P. O. Box 12194, Research Triangle Park, N. C. 27709

I. INTRODUCTION

The transport of water through a hydrophilic polymer and the molecular mobility of water in such polymers are considered in this study using systems that are as nearly homogeneous as can experimentally be achieved; these are hydrogels, which offer the added advantage that they can be prepared with widely varying water contents. Transport properties and molecular mobility can therefore be studied in dependence on the degree of hydration, that is the volume fraction of water in these gels. The homogeneity of the gels circumvents some of the difficulties in interpreting experimental results which are often encountered with polymeric systems having a superstructure.

Viscous flow under hydraulic pressure has been studied previously in similar hydrogels by Refojo.[1] In the present paper measurements of viscous flow as well as activated diffusion are reported as functions of hydration. These combined data suggest a vast difference in molecular mobility of the water molecules between samples of high and low hydration; stated another way, the data indicate that the average properties of water change gradually from a state in which water is tightly bound to the polymer at low hydration to a state resembling free water at high hydration.

An attempt was made to study this aspect in more detail by the experimental techniques of wide line nuclear magnetic resonance (NMR) and differential scanning calorimetry (DSC). NMR revealed appreciable large scale molecular motion in the hydrogels at subzero centrigrade temperatures, where neither pure water (ice) nor the

pure polymer show such mobility. In the hydrogel the extent of
this motion was studied in dependence of hydration, offering
additional clues concerning the state of water in these systems.
The DSC results allow one to estimate the amount of water not
available for ice formation. This figure is related to the low
temperature mobility and transport properties of the hydrogels.

II. EXPERIMENTAL

A. Materials

a) The hydrogel membranes were formed by simultaneous polymerization and cross-linking of the monomers. In order to vary
the equilibrium water content of the final membranes, copolymerization with a cross-linking agent (tetraethylene glycol dimethacrylate)
or copolymerization with a series of monomer homologues of varying
hydrophilicity (glycerol monomethacrylate, ethylene glycol monomethacrylate, hydroxy propyl monomethacrylate, methyl methacrylate)
were used. All hydrogel membranes were prepared without support
by polymerizing the monomer solution between two glass plates
separated by appropriate spacers to control the thickness.
b) Cast films for use as membranes of low water content were
prepared by casting solutions of various polymers onto glass plates
with an adjustable casting blade and then allowing the film to air
dry slowly under cover. A series of cellulose acetate samples
with varying acetyl contents was included in this type of membranes.
All membranes, both hydrogels and cast films, were equili-
brated in pure water before the experiment, and were optically
clear. Details of the preparation of membranes and of the measurement of water permeabilities can be seen in ref. 2.
c) The hydrogel samples used for NMR studies were prepared
by polymerization of a mixture of monomer (glycerol monomethacrylate
GMA) and water in 10 mm O.D. glass tubes. The water content of
the gel was controlled by the amount of water added to the monomer
mixture, and no cross-linking agent was added. The monomer mixture
was degassed and sealed under vacuum, after which the polymerization was initiated by an aqueous redox system of $K_2S_2O_8$ and
$Na_2S_2O_5$ at room temperature. Gelation of the monomer mixtures
occurred within a few hours: after standing at room temperature
over night they were then cured at elevated temperature (80°C)
for one hour. Although no cross-linker was added, all gels were
apparently cross-linked and insoluble in water, probably due to
chain transfer reactions and also to trace amounts of di- or
trimethacrylates present in the monomer. All measurements were
done with these sealed tubes without transferring the gels. The
water content was calculated from the amount of water added assuming 100% conversion of monomer to polymer.

d) Poly-GMA samples for DSC measurements were prepared in
6 mm tubes as outlined in c) above. The tubes were opened and
small sections of the material (ca. 12 mg) were quickly sealed
in aluminum DSC sample capsules and weighed. Following the
calorimetric measurements, the exact water content was obtained
by puncturing the sample capsule and observing the weight loss
after drying at 75-80°C under vacuum ($<10^{-5}$ Torr) for 60 hrs.

B. Measurement of Water Transport

The water flux J (cm^3/cm^2 · sec) through a homogeneous polymer
membrane of unit area can be given by

$$J = K \frac{\Delta p}{\Delta x} \tag{1}$$

where K is the piezometric permeability coefficient and $\Delta p/\Delta x$ is
the pressure gradient. The flux J can be considered as the sum
of a diffusive flux J_d and a flux due to viscous flow, J_f; i.e.,

$$J = J_d + J_f \tag{2}$$

The piezometric permeability coefficient K can similarly be divided
into the diffusive permeability coefficient K_d and the flow conductivity coefficient K_f; i.e.,

$$K = K_d + K_f \tag{3}$$

When the water flux is measured with radioactive water under no
pressure gradient, the flux observed corresponds to J_d. Under
such conditions, J_d' (g/cm^2 · sec) is related to the diffusive
permeability P and the concentration gradient of radioactive water
$\Delta c/\Delta x$ as

$$J_d' = P \frac{\Delta c}{\Delta x} \tag{4}$$

From the consideration of the chemical potential of water, P and
K_d can be related as[2]

$$K_d = P \frac{v_1}{RT} \tag{5}$$

where v_1 is the molar volume of water. Consequently, K is given
by

$$K = P \frac{v_1}{RT} + K_f \tag{6}$$

The measurement of K was carried out using conventional
ultrafiltration cells (Amicon Ultrafiltration Cells Model 50
and Model 420). Water flux was measured at various applied
pressures and the initial linear slope in the plot of flux versus
pressure was taken as the value of K. The pressure used for the
measurement was varied depending upon the water content of the
membranes; e.g., with highly hydrated membranes applied pressures
of 10 to 30 psi were used, but with membranes of low hydration
pressures of up to 550 psi (using Model 420) were used.

The diffusive permeability P was measured in a Lucite dialysis
cell using tritiated water. At "time 0" one side of the cell was

dosed with 50µl tritiated water (specific activity: 1.0 milli-
curie/gram). The water on both sides of the membrane was stirred
at a rate of 240 r.p.m. At subsequent time intervals small aliquots
were simultaneously withdrawn from both sides of the cell, and
their radioactivities (counts per minute) were measured by a Packard
Tri-Carb spectrometer (Unilux II).

The diffusive permeability P was calculated from the slope
of log ΔN versus time t according to the equation

$$P/\Delta x = - \frac{V}{2A} \frac{\Delta \ln \Delta N}{\Delta t} = - \frac{2.3V}{2A} \cdot \frac{\Delta \log \Delta N}{\Delta t}$$

where A is the membrane area, V is the volume of each cell half,
Δx is the membrane thickness, ΔN is the difference in "count"
between the two cell halves at time t, and Δt is the elapsed time
from any arbitarily chosen starting time. All transport measurements were made at 23-24°C.

C. Nuclear Magnetic Resonance (NMR)

Wide line NMR spectra were obtained using a Varian DP-60 spectrometer. Spectra were taken, starting at room temperature, in 5° or
10°C intervals at successively lower temperatures down to about
-100°C. This whole cooling process took approximately 8 hours.

The water content of the samples is recorded in Table I. The
fraction of hydrogens (or protons) in the sample that are attached
to water is given in the last column of this table.

TABLE I

Volume Fraction (Hydration) and Mass Fraction of Water and the
Fraction of Water Protons in the poly-GMA Samples Used for NMR

Hydration	Mass Fraction of Water	Fraction of Protons in Water
0.0	0.0	0.0
0.1	0.09	0.12
0.2	0.18	0.24
0.3	0.27	0.36
0.4	0.37	0.46
0.5	0.46	0.56
0.9	0.89	0.92

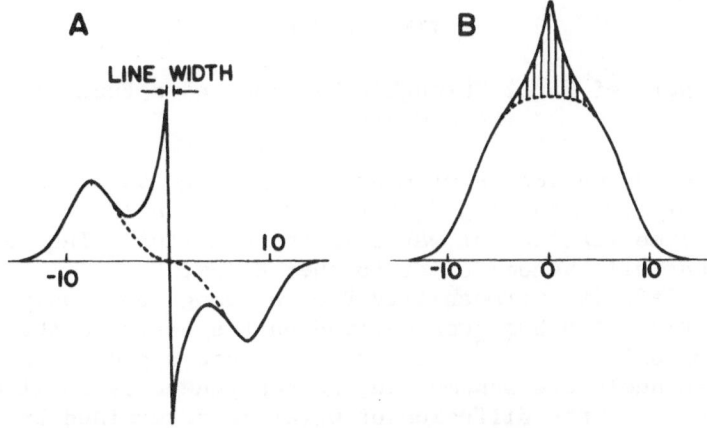

Fig. 1 a) Schematic representation of the experimentally obtained
            first derivative absorption as a function of the mag-
            netic field from the center of resonance, measured in
            Gauss. Both a broad and a narrow line are present. The
            separation of the lines is indicated.

        b) The actual absorption spectrum constructed from the
            first derivative absorption. The area under the
            narrow line is shaded.

At certain temperatures the observed first derivatives of the
proton absorption spectra have the shape schematically indicated
in Fig. 1 a. They consist of a broad line, which corresponds to
relatively immobile protons, and a narrow line, which arises from
protons on highly mobile water molecules and polymer chain segments.
Two quantities will be determined from such spectra and used in the
discussion to follow: First, the fraction of mobile protons is
estimated by taking the ratio of the area under the narrow absorption
line to the area under the whole absorption, as indicated in Fig. 1 b.
Second, the width of the narrow line is determined in the manner
shown in Fig. 1 a. The smaller the width of the narrow line, the
higher is the mobility of the molecules whose protons contribute
to that line, in general.

D. Differential Scanning Calorimetry (DSC)

DSC traces were obtained with a Perkin-Elmer DSC-1; heats of
fusion were evaluated by mechanical integration of the curves. As
broad crystallization exotherms were noticed in certain samples on
heating above -50°C, the following program was adopted: The specimen
was cooled from room temperature at 20°C/min to -90°C, then heated
at 10°/min to -20°C at which point any delayed crystallization had
taken place. After cooling again to -90°C, the final DSC run was
taken at 10°/min up to 30°C.

III.   RESULTS

A.   Transport of Water Through a Membrane as a Function of Water
Content

The measured values of K and $K_d = \dfrac{P v_1}{RT}$ are summarized in Fig. 2
as plots of log K and log K_d versus X = (1 - H)/H, where H (hydration)
is the volume fraction of water in the membrane.   The reason for
this format will become clear further below.

The diffusive permeability P of a homogeneous polymer membrane
as a function of H has been derived on the basis of the free volume
concept of diffusion.[2,3]   According to this approach, no fixed
pore or channels are assumed in the homogeneously swollen polymer
(membrane), and the diffusion of water is determined by the total
free volume.   This total free volume in the membrane consists of

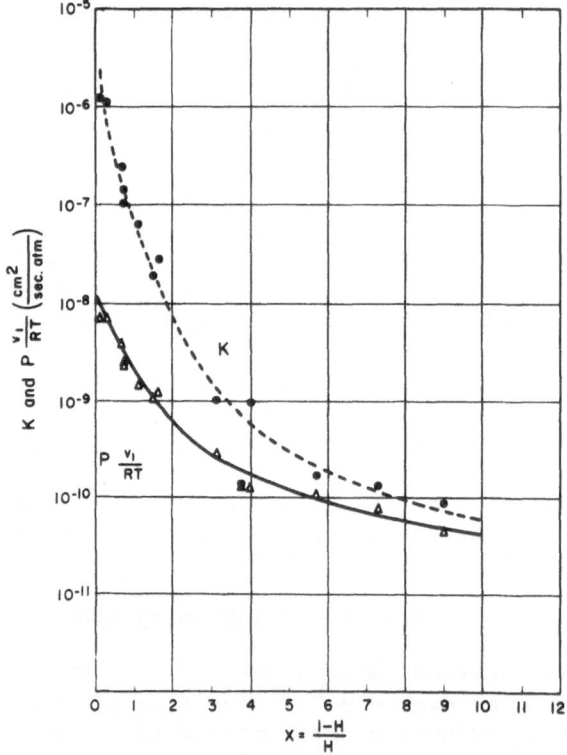

Fig. 2   The piezometric permeability coefficient K and the
         diffusive permeability coefficient K_d = P v_1/RT as a
         function of the parameter X = $\dfrac{(1-H)}{H}$, where H is the
         volume fraction of water in the water swollen (homo-
         geneous) membrane.   The solid line represents Eq. (7)
         with α = 0.5 and β = 4.5.

both the free volume of the polymer and of water. The change of
free volume can be correlated to the hydration H of the membrane.
With this model the diffusive permeability P of water in a water
swollen polymer membrane is given by

$$\ln \frac{P}{D_o} = - (1 - \alpha)\beta \frac{X}{1 + \alpha X} - \ln (1 + X) \qquad (7)$$

where α and β are constants specific to water-polymer system, D_o
is the self diffusion constant of water, and X is the parameter
used in Fig. 2 and given by

$$X = \frac{(1 - H)}{H} .$$

The details of derivation and meaning of the constants are given
in ref. 2.

The solid line for P shown in Fig. 2 is the calculated theoretical
line using values of α and β obtained from the experimental data. The
diffusive permeability P decreases from the value of D_o as the
parameter X increases (H decreases). As can be seen in Fig. 2 the
theoretical curve and experimental data agree quite well; thus
Eq. (7) describes P for the entire range of H.

The piezometric permeability coefficient K cannot be described
by Eq. (7), since the mode of water transport under applied pressure
may be completely different from that in diffusion. The results
indicate that 1) K and P v_1/RT are nearly identical in the low
hydration region (large X), 2) K and P v_1/RT are clearly different
in the high hydration region, and 3) $K_f = K - P v_1$/RT increases
with increasing water content in the high H region, the contribution
of viscous flow (given by K_f) to the piezometric permeability
coefficient K becoming dominant at very high hydration.

The dependence of K_f on H can be derived by considering the
frictional resistance caused by the macromolecules of the membrane.
In conformity with the treatment of the free volume concept of
the diffusive permeability, it is assumed that the membrane macromolecules and the permeant molecules (water) are initimately mixed.
Every polymer chain is individually opposing the flow by a viscous
resistance, thereby contributing to the flow resistance $1/K_f$.
Such a concept was introduced by P. Debye and A. M. Bueche[4] for
the treatment of intrinsic viscosity and translational diffusion
of randomly coiled macromolecules. The driving force F overcoming
the resistance to flow through 1 cm^3 of the membrane reads

$$F = \frac{dp}{dx} = nfv = J_f/K_f$$

and the viscous flow current

$$J_f = Hv ,$$

where n is the number of repeat units per cm^3, f their friction
coefficient and v is the average flow velocity. With

$$n = (1 - H) \rho N_L/M ,$$

where M is the molecular weight per repeat unit, ρ the density of the dry membrane, and N_L Avogadro's number, one obtains

$$K_f = \frac{M}{f\rho N_L} \cdot \frac{1}{X} \; . \tag{8}$$

This relation must be modified at lower hydration when there is not sufficient permeant for complete solvation of the membrane macromolecules and for flow between them. One can assume that this effect brings the viscous flow of permeant to zero at a critical hydration H_c which corresponds to the point where $K_f = 0$, and $K = K_d = P \, v_1/RT$. With this modification, Eq. (8) becomes

$$K_f = \frac{M}{f\rho N_L} \left(\frac{1}{X} - \frac{1}{X_c} \right) \; , \tag{9}$$

yielding $K_f = \infty$ at $H = 1$ ($X = 0$) and $K_f = 0$ at $H = H_c$ ($X = X_c$). The data in Fig. 2 are replotted as $K_f = K - P \, v_1/RT$ versus $H/(1 - H) = 1/X$ in Fig. 3. From this plot $1/X_c$ was estimated as 0.75 and

$$\frac{M}{f\rho N_L} = 2.88 \times 10^{-7} \; cm^2/sec \cdot atm \; .$$

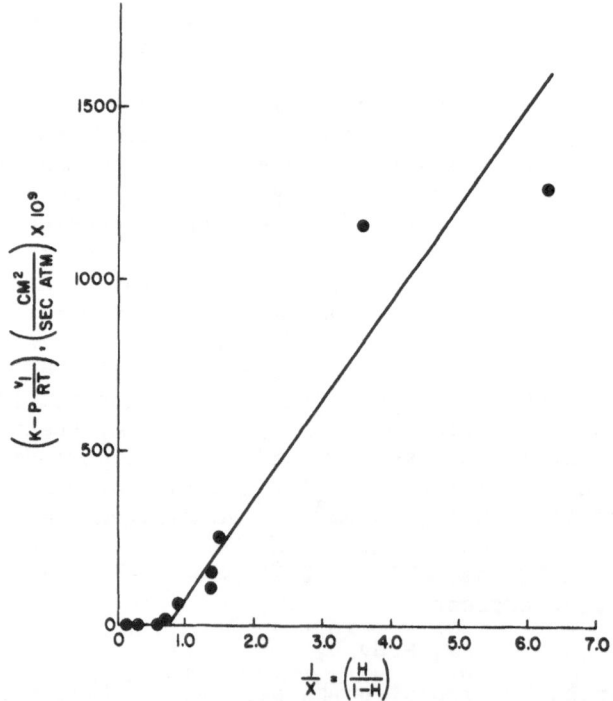

Fig. 3 The flow conductivity coefficient $K_f = K - P \, v_1/RT$ as a function of the parameter $1/X = H/(1 - H)$.

Using these values, the theoretical K_f values (solid line are compared with experimental data in Fig. 4. Evidently, the dependence of K_f on X is indeed in accordance with Eq. (9). The value of K_f asymptotically approaches infinity as X = (1 - H)/H goes to zero. In Fig. 4 this is noticeable for values of X smaller than 0.1. In the range of X values from 0.1 to 1.1 the curve can be well approximated by a straight line. This seems to be an explanation for the apparently linear dependence of log K versus X found for many hydrophilic polymer membranes.

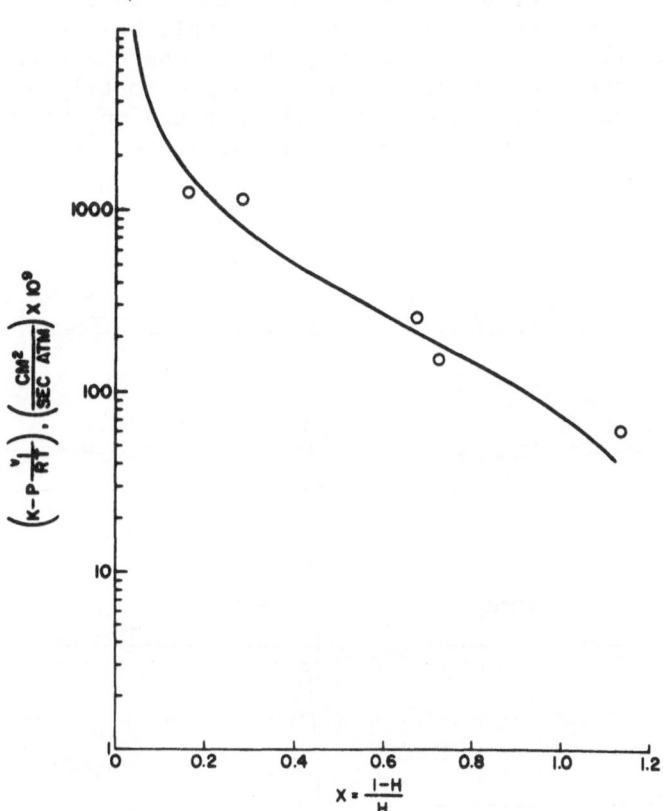

Fig. 4 The dependence of the flow conductivity coefficient
$K_f = K - P\ v_1/RT$ on the parameter X = (1 - H)/H.
The solid line represents Eq. (9), with $1/X_c$ = 0.75
and $M/f\rho N_L$ = 2.88 x 10^{-7} cm^2/sec·atm.

B. Nuclear Magnetic Resonance

Some typical spectra obtained with poly-GMA of hydration 0.1 and 0.5 are shown in Fig. 5 at various temperatures. These are the experimentally obtained first derivatives of one half of the symmetric proton absorption. It is readily apparent that with increasing temperature a narrow line develops and increases in intensity. Below -100°C the spectrum no longer changes with temperature, as the system is completely frozen at -100°C and large scale molecular motion has ceased.

In this essentially rigid lattice the magnetic dipole-dipole interaction of the protons leads to maximal broadening of the spectrum. The appearance of a narrow line at -60°C in addition to the broad line (Fig. 6), on the other hand, indicates that a certain fraction of the hydrogens in the sample has attained large scale mobility at frequencies greater than the critical frequency in these experiments, which is approximately 5×10^4 Hz.[5] This mobile fraction, estimated from the spectra in the manner described above, is shown in Fig. 6 for the different samples as a function of temperature.

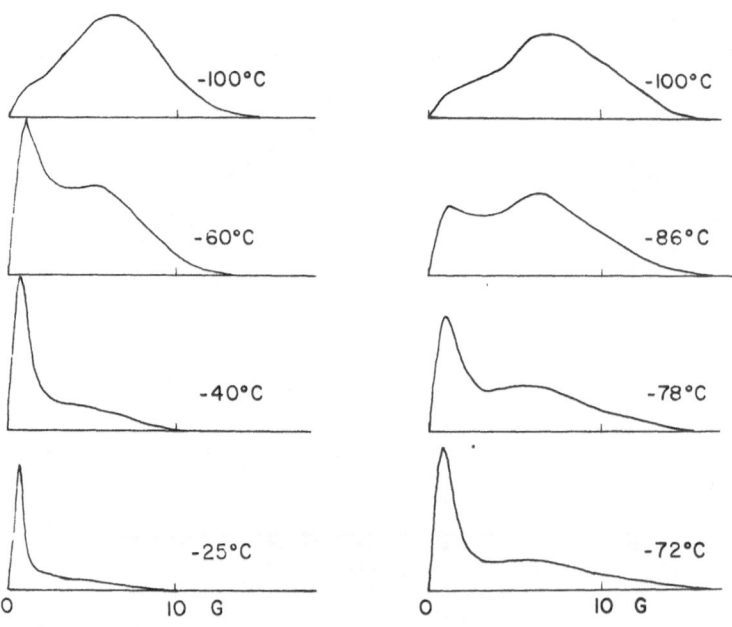

Fig. 5 The first derivative proton absorption of poly-GMA versus the magnetic field from the center of resonance. The spectra were taken at the indicated temperatures. Left: hydration 0.1; right: hydration 0.5.

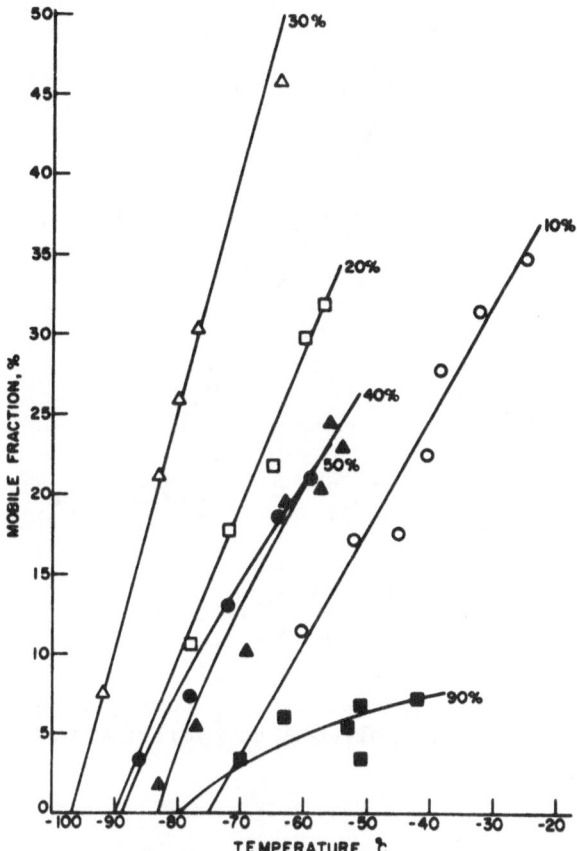

Fig. 6   The mobile fraction (fraction of mobile hydrogen) as a
function of temperature for the poly-GMA samples having
the indicated hydration.

It should be mentioned that neither the pure polymer, poly-GMA,
nor pure water (ice) show a narrow line below 0°C. It is therefore
the particular interaction between polymer and water that gives
rise to the considerable mobility observed in the mixture.
    Not only the intensity of the narrow line (mobile fraction),
but also the width of this line is strongly dependent on the
temperature as shown in Fig. 7 for poly-GMA of hydration 0.9.
Around 0°C, supercooling was noticeable which is also believed
to be responsible for the difference between the line width
observed on cooling (the circles in Fig. 7) and on warming (triangles).
The smallest line widths measured around 0°C are not considered
to be accurate but essentially to be given by the inhomogeneity

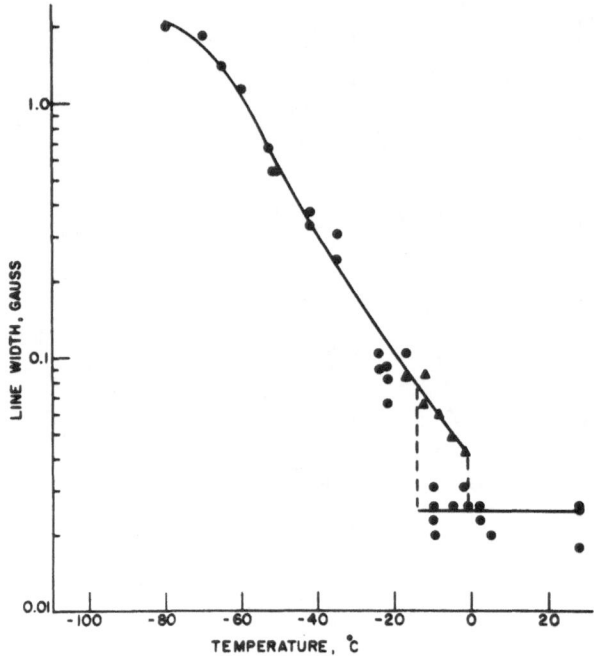

Fig. 7   The width of the narrow line, in Gauss, as a function
of temperature for poly-GMA at hydration 0.9.   See
text.

of the magnetic field.

It should be noted that the mobile fraction in Fig. 6 does
not simply represent the fraction of water unfrozen at a given
temperature.   For example, the fraction of water protons in the
hydrogel with hydration 0.1 is 0.12 (Table I), whereas the fraction
of mobile hydrogens is equal to 0.35 at -25°C.   This implies that
hydrogens on polymer molecules contribute substantially to the
mobile fraction; both water and polymer molecules are mobile at
temperatures below 0°C.   Unfortunately, it is impossible to
determine the fraction of mobile water and the fraction of mobile
polymer separately from the present data.

The striking feature of the curves in Fig. 6 is that the
samples with hydration 0.1, 0.2, and 0.3 show a consistent shift
of their mobile fraction versus temperature curves toward lower
temperatures, the curves also becoming steeper.   This trend is
reversed with samples of hydration higher than 0.3.   Thus the
sample with 0.4 hydration is shifted to higher temperatures with
respect to the sample with hydration 0.3.

The following explanation is offered for this behavior. When initially clear hydrogels with hydration greater than 0.3 are cooled to low temperatures one invariably observes that the samples become turbid. It is well known that this turbidity is caused by ice formation. Hydrogels with hydration 0.3 or less, however, remain clear on cooling. In this case, ice formation does not take place. This observation agrees with reports on gelatin-water mixtures, stating that water does not crystallize on cooling when present in concentrations of 35 weight -% or less.[6,7] That water is not available for ice formation in this concentration range is interpreted as due to interaction between water and the polymer.[6] In the hydrogels water seems to act as a plasticizer at hydrations below 0.3; the mentioned shift of the curves to lower temperatures (Fig. 6) may accordingly be interpreted as a lowering of the glass transition temperature.

The absence of ice formation at hydrations below 0.3 has been attributed above to water-polymer interaction; since it could be argued that other factors are responsible, this contention will be substantiated further. As an alternative explanation one could suggest that at low hydration the separation of water molecules and lack of migration prevent ice formation. This argument, however, does not apply at hydrations between 0.2 and 0.3, for instance. First, the separation of water molecules is only a few Ångstrom units, and second, the NMR data make it clear that mobility at subzero temperatures is certainly high enough to ensure migration.

When ice formation does occur at hydrations above 0.3 it is difficult to account for the behavior of the curves in Fig. 6 in detail. Qualitatively one can attribute the shift of the curves to higher temperatures to the following two reasons. First, the crystallizing water is lost for plasticization, and second, the presence of ice presumably imposes a hindrance to polymer mobility.

C. Differential Scanning Calorimetry

All the poly-GMA samples having hydrations greater than 0.3 displayed rather broad DSC endotherms starting around -15°C; a sample of 0.23 hydration had no observable transition between -90° and room temperature. The DSC traces are not reproduced here, as the large sample sizes and high heating rates lead to super-heating effects which make the shape and absolute temperature of the peaks of questionable value.

Of primary interest to the present study are the heats of fusion of the various hydrogel preparations; these are summarized in Table II. To quantitatively examine the notion that a certain fraction of the water is closely associated with the polymer molecule and therefore does not crystallize, we have plotted the data in the last two columns of Table II according to the expression

$$\Delta H \left[\frac{cal}{g \text{ dry polymer}}\right] = \Delta H_f (W_{tot} - W_{unf}) \quad . \quad (10)$$

Here ΔH_f is the heat of fusion of the water able to freeze and melt; W_{tot} and W_{unf} are the weight of total water and nonfreezing water, respectively, per gram of dry polymer. It is apparent from the experimental points in Fig. 8 that the observed heat of

TABLE II

Heat of Fusion of Water in poly-GMA Hydrogels

Hydration	W_{tot} $\dfrac{g\ H_2O}{g\ dry\ polymer}$	ΔH_{obs} $\dfrac{cal}{g\ sample}$	ΔH $\dfrac{cal}{g\ dry\ polymer}$
0.23	0.264	0.0	0.0
0.34	0.460	5.7	8.3
0.36	0.495	8.2	12.3
0.47	0.745	17.5	30.6
0.63	1.528	35.4	88.7
0.75	2.777	50.7	191.3

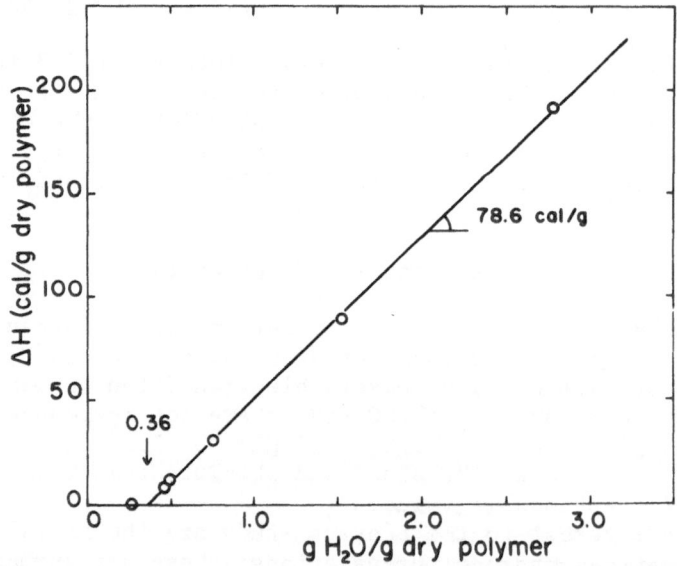

Fig. 8   Heat of fusion as a function of W_{tot} (g H_2O/g dry polymer) for the poly-GMA hydrogels. The solid line represents Eq. (10) with ΔH_f = 78.6 cal/g H_2O and W_{unf} = 0.36.

transition is a linear function of the quantity of crystallizable water ($W_{tot} - W_{unf} \geq 0$) present. A least-squares fit of the data (solid line in Fig. 8) yields $\Delta H = 78.6$ cal/g, in good agreement with the normal heat of fusion of water (79.7 cal/g), and $W_{unf} = 0.36$ g unfrozen H_2O/g dry polymer. It should be noted that the water in these hydrogels does not exhibit the anomalous "fusion-like" transitions found by Haly and Snaith in the keratin-water system.[8] The present results appear similar to those found by Dehl on water-swollen collagen fibers,[9] in that a constant amount of water does not freeze, and any other water present does crystallize and melt with the normal heat of fusion.

From the foregoing analysis we can conclude that any water present in these poly-GMA hydrogels in excess of 0.36 g/g dry polymer crystallizes and melts with the normal heat of fusion. The amount of water that will not crystallize corresponds to a volume fraction of 0.29, a value in the hydration range at which the transport properties and low-temperature mobility of the hydrogels undergo significant changes.

IV. DISCUSSION

From the results obtained by three types of experiments, one can draw certain common and consistent conclusions regarding the state of water in the hydrogels. The transport properties indicate that water moves through a swollen polymer of low hydration exclusively by activated diffusion, whereas in a highly hydrated polymer viscous flow can be observed under pressure. This implies that water exists in a molecularly dispersed form in polymers of low hydration, whereas bulk water exists at higher hydrations.

The results of differential scanning calorimetry are in good agreement with this description. DSC shows that water does not crystallize in a hydrogel of hydration less than 0.29; at greater hydrations, however, crystalline water has been detected and was shown to have a heat of fusion virtually identical to that of pure ice. This study also indicates that a constant amount of water, 0.36g H_2O/g polymer does not participate in ice formation, implying that this water is rather strongly associated with the polymer throughout the entire hydration range.

According to these DSC results it is tempting to suggest that water can be divided into "free water" and "bound water". The molecular motion observed by NMR, at subzero temperatures, however, clearly shows that both the water and the macromolecular segments are highly mobile even in the low hydration region. Consequently the word "bound water" may cause considerable misunderstanding of the true state of such water. From results of this study, one can conclude that part of the water in such water swollen polymers is highly associated with the macromolecules but it retains substantial mobility. It is realized that the exact amount of water

which can be closely associated with a particular polymer will
depend on hydrophilicity, steric factors, etc. However, the basic
trends found in this study may apply to all water swollen polymers,
so long as a system does not become obviously heterogeneous at
room temperature.

One final point regarding the results of the three types of
measurements should be mentioned. The NMR and DSC studies show
that no water is available for ice formation below H = 0.29
(Figs. 6 and 8). The viscous flow conductivity coefficient K_f,
on the other hand, vanishes at $1/X_c$ = 0.75 or H = 0.43. Thus
it is apparent that even though some water is able to crystallize
in the region H = 0.3 - 0.4, the viscous flow of this water under
a pressure gradient is unmeasurably small. This situation might
be attributed to one of two causes: 1) the transport measurements
were made on a series of hydrogels with possibly varying "bound
water" concentrations, and 2) the first water added which is able
to freeze does not solvate the polymer segments sufficiently to
permit viscous flow.

The subject of this study, hydrogels in which water and polymer
are initimately mixed, can be thought of as a homogeneous model
system for the consideration of water and hydrophilic macromolecules
in general. In particular the present results demonstrate that
a well-defined quantity of labile water is closely associated
with the polymer chain segments at all observed hydrations. This
strongly suggests that the interface between a hydrophilic polymer
and water consists of a considerable fraction of "hydrated" segments.

SUMMARY

Transport of water in water-swollen polymers can be character-
ized by the diffusive permeability P (measured by the diffusion of
tritiated water) and the hydraulic permeability K (measured by
water flux under hydraulic gradient). In low hydration region,
water moves by diffusion even under an applied hydraulic pressure
gradient. In high hydration region, it moves partly by bulk flow
under applied hydraulic pressure gradient, but moves by diffusion
only if there is no hydraulic pressure gradient. Change of mode
of transport occurs as a gradual transition as hydration increases.
Mobility of water is examined by NMR and differential scanning
calorimetry as function of temperature. Mobility below the freezing point of water and freezing phenomena observed by differential
scanning calorimetry indicate that change of water mobility in
swollen polymers occurs as gradual transition as hydration varies.
These observations suggest that water in swollen polymers is different from free water; no distinct state of water can yet be
assigned to it. Water seems to change gradually during increasing
interaction with polymer molecules.

References

1. M. F. Refojo, J. Appl. Polymer Sci., 9, 3417 (1965).
2. H. Yasuda, C. E. Lamaze and A. Peterlin, J. Polymer Sci. A2, in press.
3. H. Yasuda, A. Peterlin, C. K. Colton, K. A. Smith and E. W. Merrill, Makromol. Chem., 126, 177 (1969).
4. P. Debye and A. M. Bueche, J. Chem. Phys. 16, 573 (1948).
5. E. R. Andrew, Nuclear Magnetic Resonance, Cambridge University Press, 1958.
6. T. Moran, Proc. Royal Soc., 112A, 30 (1926).
7. B. J. Luyet, Annals New York Academy Sci., 125, 502 (1965).
8. A. R. Haly and J. W. Snaith, Biopolymers, 7, 459 (1969).
9. R. E. Dehl, Science 170, 738 (1970).

Kinetic and Spectral Evidence for Selective Solvation of a Hydrophobic Quinoneimine Dye in Mixed Aqueous Solvents

R. L. Reeves and R. S. Kaiser

Research Laboratories, Eastman Kodak Company

Rochester, New York 14650

Water-soluble dyes are useful probes for the investigation of changes in solvent structure around large molecules. Their size bridges the gap between macromolecules and small ions, whose solvation has been studied extensively. Most of the bulk of a dye molecule is hydrophobic and capable of acting as a "structure-former" in water, although the orientation of water molecules in the hydration sphere may not be as uniform as might be expected for a hydrocarbon solute of similar bulk. Dyes offer a further advantage in having well-resolved absorption bands that are sensitive to changes in solvent polarity.[1-3]

We found recently that azo dyes which exist in two tautomeric forms show a pronounced shift in the tautomeric equilibrium when organic solvents are added in small amounts to aqueous solutions of the dyes.[4] The greatest shift in the equilibrium occurs in the predominantly aqueous compositions where the solvent retains three-dimensional structure, and we interpreted the results in terms of selective solvation of the dyes by the organic solvent within the water structure.

The exchange of a water-solute interface by an organic solvent-solute interface in a predominantly aqueous medium should have a pronounced effect on the kinetics of reactions of a hydrophobic solute. The effect should be different for bimolecular and unimolecular reactions. Observation of a dramatic change in rate upon addition of small amounts of organic solvent to an aqueous medium that could not be accounted for in terms of changes in bulk solvent properties would indicate that selective solvation of the solute or of the activated complex, or both, was occurring.

A kinetic study of the alkaline hydrolysis of the naphthoquinoneimine dye (1) provides a unique opportunity to seek
kinetic effects of selective solvation. Two ionic sites lie
along the rim of the planar molecule, but most of the bulk is
hydrophobic. The hydrolysis consists of two steps involving
one bimolecular and two quite different unimolecular reactions.[5,6]
Unlike the hydrolysis of most aryl imines, the intermediate
carbinolamine (3) accumulates, and the three rate constants can

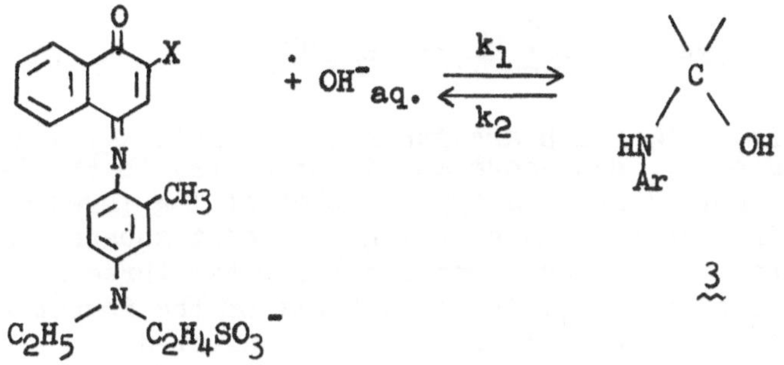

1, X = CONHC$_2$H$_4$SO$_3$$^-$

2, X = CON(C$_2$H$_5$)$_2$

(1)

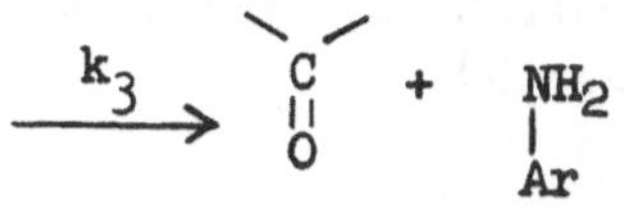

each be evaluated from single kinetic runs. Since the reaction
site is in the hydrophobic moiety, the free energy of activation
for formation of 3 will contain contributions from the energy
required for the hydrated OH$^-$ to pass through the solvation
sphere and to begin to shed its waters of hydration. The activation energy for conversion of 3 to products will contain
contributions from solvation energies of the bulky intermediate,
the activated complex, and the two product ions.

Kinetic measurements were made at 25° in binary mixtures of
water with methyl, ethyl, isopropyl, and tert-butyl alcohols and
dimethyl sulfoxide (DMSO) at an ionic strength of 0.1.
Measurements could be made over the entire range of compositions
in the methanol- and ethanol-water mixtures. Miscibility limits
were reached above 8 mol % t-BuOH, and above 50 mol % DMSO and
i-PrOH in the presence of the sodium hydroxide-sodium chloride
electrolyte.

The accumulation of intermediate could usually be controlled by adjustment of the base concentration in the range 0.01 to 0.1 M. The rapid disappearance of dye under these conditions was followed on a Durrum stopped-flow spectrophotometer. In water and in aqueous solutions containing 2-5 mol % organic solvent, k_2, k_3, and k_1' ($=k_1[OH^-]$) were of comparable magnitude, and the [dye] -time plot had a fast and a slow component. The solution to the rate equation has the form of eq (2) when [OH^-] is constant throughout a run.

$$[Dye] = B_1 e^{-\alpha t} + B_2 e^{-\beta t} \qquad (2)$$

The parameters B_1, B_2, α, and β are functions of k_1', k_2, and k_3.[5] They were evaluated by least-squares fit of eq (2) to the data for each run. Figure 1 shows a typical plot of log [Dye] against t. The solid line was calculated using the least-squares values of the parameters. Three equations involving the three k's in various combinations can be written in terms of the four parameters, from which the individual k's were evaluated.

At mole fractions of organic solvent greater than about 0.05, k_3 decreased relative to k_2 to where the equilibrium giving the intermediate was established. Under these conditions the kinetics were treated separately as a rapid first-order approach to equilibrium followed by a slower first-order decay of dye from the equilibrium concentration. The three rate constants were evaluated from eqs (3)-(5):

$$k_{OBS} \text{ (first stage)} = k_1' + k_2 \qquad (3)$$

$$K = \frac{[Intermediate]}{[Dye]\;[OH^-]} = \frac{A_o - A_e}{A_e\;[OH^-]} = \frac{k_1}{k_2} \qquad (4)$$

$$k_{OBS} \text{ (second stage)} = k_3\,K\;[OH^-] \qquad (5)$$

where the A's are dye absorbances and the subscripts refer to initial and equilibrium values.

In a kinetic run, a solution of dye in solvent was mixed with a mixture of NaOH and NaCl in solvent in the stopped-flow spectrophotometer. For those runs where the second stage of the reaction was much slower than the first, a stop watch was started at the instant the solutions were mixed. The time-base of the

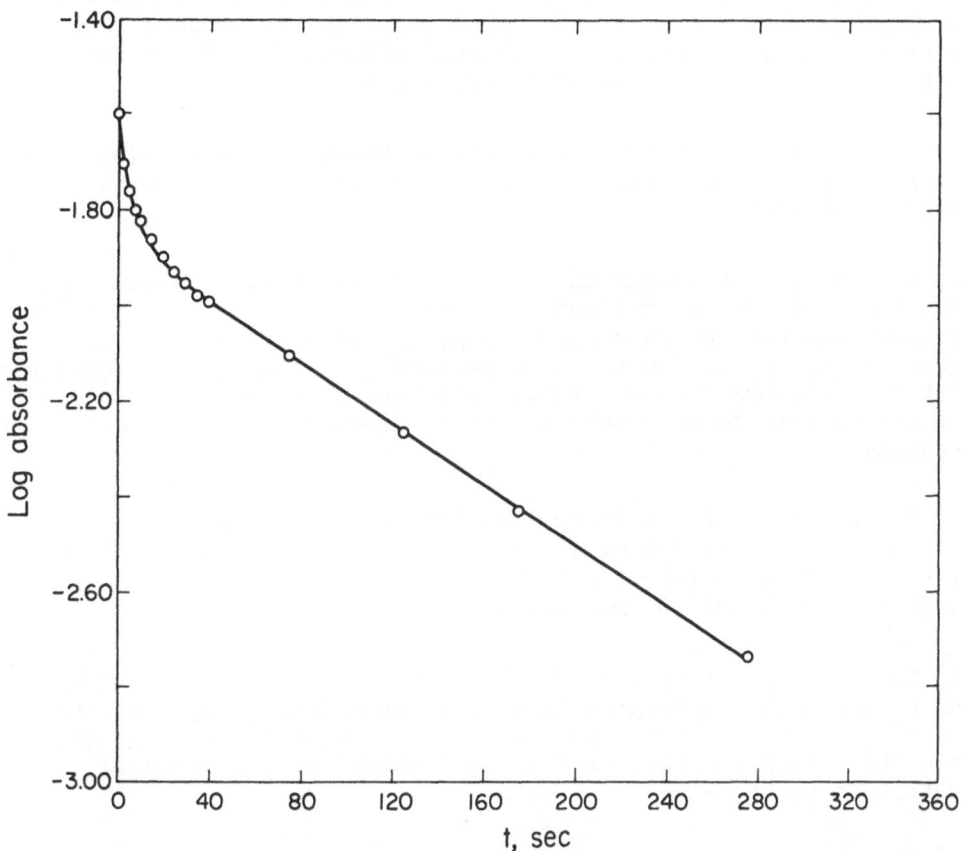

Figure 1.   Plot of log absorbance (680 nm) vs. time for dye 1
            in 0.005 M NaOH and 0.095 M NaCl in water at 25°.

oscilloscope was used for the initial rapid reaction. The
oscilloscope trace was triggered manually at intervals timed by
the stop watch throughout the slow stage. The combined traces
were recorded together on the storage screen and photographed.

    Absorption curves in the visual region were obtained for the
entire range of compositions in mixtures of water with methyl,
ethyl, isopropyl, and tert-butyl alcohols and in dimethylformamide
(DMF) and DMSO. The dye concentrations were 2.5 - 3.5 x 10^{-6} M,
a range where the solutions obeyed Beer's law. The solutions
contained no electrolyte and were prepared by diluting a stock
solution in water with the mixed solvent. The composition of
solvent in the final solutions was calculated from the volumes
mixed and from the published densities of the binary mixtures.

The variation in the three rate constants with solvent composition is shown in Figure 2. Each point is the average from several runs at different base concentrations. The following points about the plots should be stressed:

(1) In nearly every case, the greatest change in rate constant occurs when small amounts of organic solvent are added to the aqueous solutions.

(2) Nearly all the change in the unimolecular rate constant, k_2, occurs at solvent compositions between zero and 0.02 mole fraction methanol and between zero and 0.05 mole fraction ethanol. In these ranges, k_2 is increased by factors of 40 and 13, respectively. At higher alcohol concentrations, the rate at which the intermediate reverts to reactants is nearly insensitive to solvent changes.

(3) The greatest change in the bimolecular rate constant, k_1, occurs between zero and 0.07 mole fraction ethanol and between zero and 0.15 mole fraction methanol. The values of k_1 increase by factors of 7 and 50, respectively.

(4) The order of effectiveness of the alcohols in increasing k_1 and k_2 and in decreasing k_3 is methyl > ethyl > isopropyl > or tert-butyl. The two bulky alcohols cause an initial decrease in k_1. This order suggests that steric requirements of the alcohols contribute to the kinetic effects.

(5) The changes in rate constants do not parallel the changes in solvent dielectric constant of the mixed solvents.

(6) Throughout the whole range of solvent compositions, the kinetic effects of the various organic cosolvents are specific for each solvent.

Table I lists values of ΔF for conversion of reactants to carbinolamine in the various solvent combinations ($K = k_1/k_2$). Also listed are changes in ΔF ($\Delta \Delta F$) for the equilibrium on transfer from water to mixed solvent. For comparison, changes in the free energy of activation for the formation of intermediate ($\Delta \Delta F_1^{\ddagger}$) and for reversion of intermediate to reactants ($\Delta \Delta F_2^{\ddagger}$) on transfer from water to 0.1 mole fraction cosolvent in water are shown. There is negligible change in ΔF on transfer from water to 0.1 mole fraction methanol, and all the changes in k_1 and k_2 can be ascribed to a lowering of $\Delta F^{\ddagger}$. As one proceeds through the series to increasingly bulky alcohols, it is seen that $\Delta \Delta F$ for transfer from water to 0.1 mole fraction alcohol becomes increasingly positive, and the free energy of activation shows a corresponding increase.

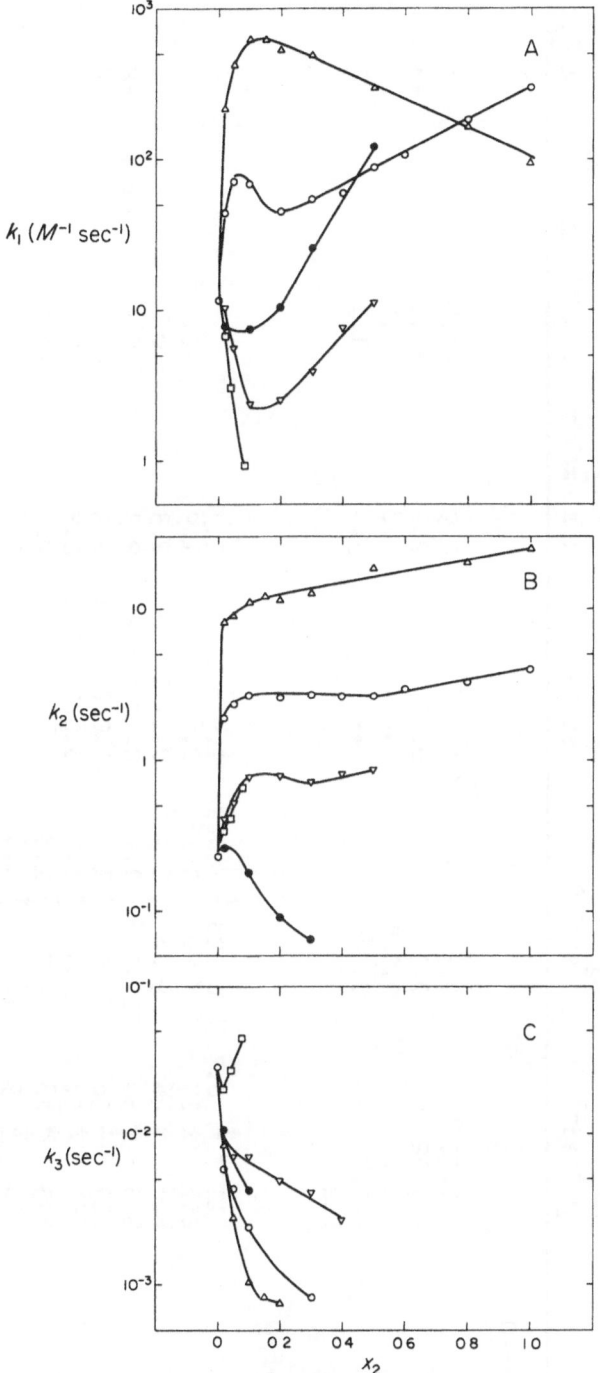

Figure 2. Semilog plots of hydrolysis rate constants of dye 1 at 25° vs. mole fraction organic cosolvent. △, MeOH; ○, EtOH; ▽, iso-PrOH; □, tert-BuOH; ●, DMSO.

Table I

Rate Data for the Hydrolysis of Dye $\underset{\sim}{1}$ at 25°, $\mu=0.1$

x_2	$k_1 \pm s$, $\underline{M}^{-1}$ sec^{-1}	$k_2 \pm s$, sec^{-1}	$10^2 k_3 \pm s$, sec^{-1}	$-\Delta \underline{F}$, kcal-mol^{-1}	$\Delta\Delta \underline{F}$	$\Delta\Delta \underline{F}^{\ddagger}_1$	$\Delta\Delta \underline{F}^{\ddagger}_2$
DMSO–H$_2$O							
0.02	7.90	0.30	1.1	1.9	+0.4		
0.10	7.52	0.19	0.45	2.2	+0.1	+0.2	+0.1
0.20	10.4	0.085	--	2.8	-0.5		
0.30	21.8	0.065	--	3.4	-1.1		
0.50	122						
MeOH–H$_2$O							
0.0	11.7 $\pm$ 0.3	0.228 $\pm$ .022	2.9 $\pm$ 1.1	2.3			
0.02	211 $\pm$ 17	8.28 $\pm$ 1.16	0.84	1.9	+0.4		
0.05	426 $\pm$ 46	9.04 $\pm$ 0.98	0.27 $\pm$ .10	2.3	~0		
0.10	625 $\pm$ 54	12.0 $\pm$ 1.4	0.10 $\pm$ .02	2.3	~0	-2.4	-2.3
0.15	619 $\pm$ 36	12.0 $\pm$ 1.8	0.082 $\pm$ .004	2.3	~0		
0.20	528 $\pm$ 54	11.3 $\pm$ 1.6	0.075 $\pm$ .011	2.3	~0		
0.30	495 $\pm$ 48	12.7 $\pm$ 1.8		2.2	+0.1		
0.50	301 $\pm$ 9	18.3 $\pm$ 0.6		1.8	+0.5		
0.80	164 $\pm$ 25	20.4 $\pm$ 2.4		1.2	+1.1		
1.00	98 $\pm$ 18	25.6 $\pm$ 4.5		0.8	+1.5		

Table I (continued)

x_2	$k_1 \pm s$ $\underline{M}^{-1}sec^{-1}$	$k_2 \pm s$ sec^{-1}	$10\,k_3 \pm s$ sec^{-1}	$-\Delta\underline{F}$ kcal-mol^{-1}	$\Delta\Delta\underline{F}$	$\Delta\Delta\underline{F}_1^{\ddagger}$	$\Delta\Delta\underline{F}_2^{\ddagger}$
EtOH-H$_2$O							
0.02	44.6 ± 1.8	1.90 ± 0.04	0.58 ± .24	1.9	+0.4		
0.05	72.2 ± 6.2	2.34 ± .60	0.43 ± .10	2.0	+0.3		
0.10	68.7 ± 6.9	2.72 ± .31	0.24 ± .07	1.9	+0.4	-1.0	-1.5
0.20	45.6 ± .5	2.59 ± .22	--	1.7	+0.6		
0.30	54.5 ± 8.1	2.70 ± .09	0.082 ± .021	1.8	+0.5		
0.40	60.4 ± 5.1	2.67 ± .34		1.8	+0.5		
0.50	88.5 ± 4.1	2.66 ± .22		2.1	+0.2		
0.60	106 ± 21	2.96 ± .39		2.1	+0.2		
0.80	184 ± 19	3.26 ± .88		2.4	-0.1		
1.00	301 ± 29	3.97 ± .78		2.6	-0.3		
iso-PrOH-H$_2$O							
0.02	10.25 ± .60	0.42 ± .06	--	1.8	+0.5		
0.05	5.63 ± .24	0.52 ± .04	0.70 ± .10	1.4	+0.9		
0.10	2.42 ± .34	0.77 ± .18	0.70 ± .16	0.7	+0.6	+0.9	-0.7
0.20	2.56 ± .20	0.77 ± .03	0.49 ± .07	0.8	+0.5		
0.30	3.97 ± .15	0.71 ± .08	0.41 ± .15	1.0	+1.3		
0.40	7.68 ± .08	0.81 ± .08	0.26 ± .07	1.3	+1.0		
0.50	11.2	0.86	--	1.3	+1.0		
tert-BuOH-H$_2$O							
0.02	6.81	0.34	2.0	1.8	+0.5		
0.04	3.03	0.41	2.7	1.2	+1.1		
0.08	0.93	0.68	4.4	0.2	+2.1	+1.5	+1.1

Examination of the dye absorption curves in the solvent mixtures gives insight into the effect of changing solvent composition on the dye alone, independently of the effect on the solvation of OH⁻ and the various transition states. Figure 3 shows the integrated intensity of the absorption band and the frequency of the absorption maximum plotted against mole fraction cosolvent. The initial addition of organic solvent gives in every case a pronounced increase in intensity and a shift in the absorption maximum to lower frequency. The effect of DMF (not shown) is initially similar to the effects of tert-butyl and isopropyl alcohols. At some solvent composition that is specific for each organic component, the direction of the frequency shift is reversed and the intensity becomes nearly constant. The initial increase in intensity is due in part to a narrowing of the absorption band so that the extinction coefficient at the frequency of the maximum shows a more pronounced increase than does the integrated intensity.

The similar spectral changes indicate similar changes in dye solvation with all the organic solvents. The fact that protic and aprotic solvents give similar spectral changes rules out specific intermolecular hydrogen-bonded interactions between dye and added organic solvent as the origin of the spectral changes. We also found similar solvent effects on the absorption curve of dye 2, which cannot form an intramolecular hydrogen bridge between the quinone carbonyl and the amide N-H group. It appears, therefore, that the solvation by added organic cosolvent is non-specific, and that the spectral changes result from the change in environment associated with replacement of the dye-water interface with a dye-solvent interface.

It is significant that the major spectral changes occur in the predominantly aqueous solvent compositions where the most dramatic rate changes are found. It is reasonable to conclude that both phenomena are related to pronounced solvation changes and that these occur when the medium is still largely "aqueous." In discussing these changes, it is convenient to divide the water-solvent compositions roughly into two regions: the first lies between 0 and 0.1 - 0.2 mole fraction cosolvent and the second lies in the higher organic compositions. Figures 2 and 3 show that these are approximately the solvent compositions where both the nature and the extent of the spectral and kinetic effects undergo a change. In the case of the alcohol-water mixtures, this region of compositions is also the region where changes in the three-dimensional structure of the bulk solvent are believed to occur.[7-9] The large kinetic changes found in this study occur in the solvent compositions where three-dimensional "aqueous" structure remains intact or is enhanced by added alcohol. Our results suggest that in these predominantly aqueous compositions the hydrophobic moiety of the dye is

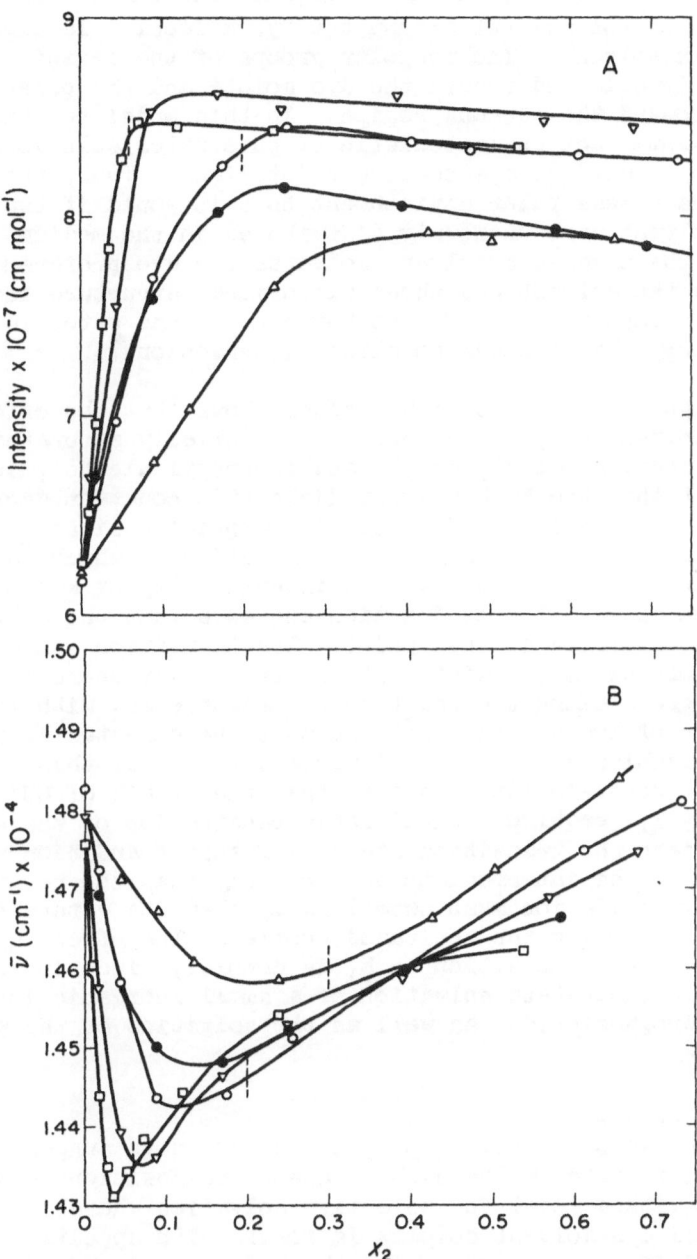

Figure 3.   Plots of absorption band intensity (A) and frequency
of maximum absorption (B) of dye $\underline{1}$ against mole
fraction organic cosolvent. $\triangle$, $\widetilde{M}$eOH; $\bigcirc$, EtOH;
$\triangledown$, iso-PrOH; $\blacksquare$, $\underline{\text{tert}}$-BuOH; $\bullet$, DMSO.

selectively solvated by the added organic solvent so that the concentration of the solvent around the dye molecules is higher than
in the bulk solvent. The nonpolar groups of the organic solvent
are probably oriented toward the dye solute and the polar
moieties toward the aqueous region. In this model the structured
dye-water interface characteristic of pure water solutions is
replaced by a diffuse dye-cosolvent interface. Thus, the dye
experiences a less polar environment that it would if the added
organic solvent were uniformly distributed in the medium. The
fact that the organic cosolvent solvates the dye preferentially
as long as the solvent has three-dimensional structure indicates
that the hydrophobic interaction between dye and water is of
higher energy than the dye-cosolvent interaction.*

In Figure 2 the plot of k_2 values shows that the effect of
changing solvent composition becomes essentially saturated at
0.1 mole fraction methyl, ethyl, and isopropyl alcohol, in spite
of the fact that the bulk solvent dielectric constant decreases
as organic solvent is added. This is surprising since k_2
measures the activation energy for a reaction in which the net
charge increases by one unit. The insensitivity of k_2 to solvent
composition above 0.1 mole fraction suggests that the solvation
of dye becomes fixed at this point. The transition state for
breakdown of the intermediate into reactants may be reached with
little progress along the reaction co-ordinate and with little
penetration of the solvent shell so that the solvated activated
complex resembles the solvated intermediate.[10] If this is correct,
microscopic reversibility requires that the attack of OH⁻,
measured by k_1, requires considerable penetration of the solvation
shell to reach the transition state. If this transition state
does resemble the intermediate more closely than the reactants,
then the hydroxide ion must have largely shed its waters of
hydration in forming the activated complex. The effect of
changing solvent composition on k_1 is complex, since it involves
changes in ground-state solvation of a small inorganic ion and
a large hydrophobic ion, as well as the solvation of the activated
complex.

---

*In the discussion of this paper, Dr. H. Friedman suggested that
the kinetic results in the highly aqueous compositions might be
consistent with a model for selective solvation whereby a
specific 1:1 dye-solvent complex is formed with specific rate
constants different from those of the hydrate. Such a model
requires a 1st-order dependence of the rates on the concentration
of added solvent. Our present data are too sparse in the low
organic compositions to test this suggestion. The occurence of
isosbestic points in the absorption curves in the high water
compositions, however, are consistent with the alternate model.

The dye absorption is the result of an intramolecular charge-transfer transition whereby excitation gives a net migration of charge from the dialkylamino group to the quinone carbonyl oxygen. We find experimentally that factors that lower the energy for the electronic transition facilitate alkaline hydrolysis. For example, the absorption maximum of dye 1 in water is shifted to lower frequencies by 840 cm^{-1}, relative to that of dye 2, and the alkaline hydrolysis rate is 2.2×10^3 times faster at a given pH. The red shift in the absorption maximum of 1 resulting from the addition of organic solvents to aqueous solutions indicates that the initial selective ground-state solvation of the dye facilitates the polarized electronic transition, and should facilitate attack by hydroxide ion. The fact that k_1 is increased initially by addition of methyl and ethyl alcohol and decreased by isopropyl and tert-butyl alcohol suggests that the internal polarization tending to facilitate attack by hydroxide ion is overridden by steric blocking of the reaction site by the bulkier alcohols as they accumulate in the solvation shell. The maximum in the plot of k_1 vs. x_2 for ethanol at 0.07 mole fraction is probably the result of steric effects taking over from polar effects as the amount of alcohol in the solvation shell increases.

The compositions greater than $x_2 = 0.1$ to 0.2 are those in which the added alcohol breaks down the characteristic three-dimensional "aqueous" solvent structure. The critical composition varies from about 0.05 mole fraction for tert-butyl to about 0.3 for methyl alcohol. The fact that extrema occur in plots of rate constants and spectral properties vs. x_2 at x_2 values between 0.1 and 0.2 also supports the idea that the medium is experiencing a pronounced change in these compositions. The blue shifts in the dye absorption maxima that we observe at x_2 values higher than 0.1-0.2 can be interpreted in terms of decreasing bulk solvent dielectric constant. These higher organic compositions are less interesting for the study of hydrophobic interactions. Never-theless, it is seen that pronounced and specific changes in k_1 occur in the less structured organic compositions. These changes are probably associated largely with changes in hydroxide ion activity. In every binary mixture where measurements could be made, k_1 either increased or decreased exponentially with mole fraction organic solvent.

The spectral data indicate that DMSO selectively solvates the dye in the predominantly aqueous compositions, but this solvation does not produce the dramatic rate changes found for the alcohols. The kinetic results indicate that in the compositions where the alcohols have the greatest effect on $\Delta F_1^\ddagger$ and $\Delta F_2^\ddagger$, DMSO has the least. Apparently, selective solvation by DMSO lowers the ground-state and transition-state energies by similar amounts. The differences between the protic and dipolar aprotic organic solvents may be due in part to differences in the

competitive hydration-solvation of the hydroxide ion in the ground
and transition states.

The lowering of the transition energy and the increase in
transition probability obtained on addition of organic co-solvent
suggests that the structured aqueous solvation shell may impose
packing or orientational strain on the dye solute which is
relieved by addition of organic solvent to the interface. There
is thus an analogy with surface tension effects whereby cosolvent will accumulate in the water interface when such accumulation decreases the surface tension. Similarly, the aqueous
solvation shell imposes an activation barrier for reaction with
hydroxide ion which may be raised or lowered by the accumulation
of organic co-solvent in the interface.

SUMMARY

The three rate constants for the alkaline hydrolysis of a
water-soluble naphthoquinoneimine dye (mol wt 662) were evaluated
in aqueous methanol, ethanol, isopropanol, t-butanol, and dimethyl-
sulfoxide (DMSO):

$$\diagup\!\!\!\!\diagdown C = N\text{-}Ar + OH^- \underset{k_2}{\overset{k_1}{\rightleftarrows}} \diagup\!\!\!\!\diagdown C \overset{OH}{\underset{NHAr}{\diagup\diagdown}} \overset{k_3}{\rightarrow} \diagup\!\!\!\!\diagdown C = O + H_2NAr$$

The variation in rate constants with solvent composition is complex. The most pronounced changes occurred in aqueous compositions where the solvent is structured. Addition of small amounts
of both protic and aprotic co-solvents increases k_1 and k_2 and
decreases k_3. The rate constant, k_1, is increased initially by
MeOH and EtOH and decreased by i-PrOH and t-BuOH. Results are
interpreted in terms of selective solvation of the hydrophobic
dye moiety by organic co-solvent. Selective solvation results
from the "structure-forming" effect of the large hydrophobic dye
on the water. Spectral data indicate that solvation by organic
solvent does not involve specific hydrogen-bonded interactions
with dye heteroatoms. Kinetic effects of selective solvation
result from opposing rate-enhancing charge redistribution and
steric blocking by bulky organic solvent.

References

1. (a) L.G.S. Brooker, G. H. Keyes, and D. W. Heseltine, J. Amer. Chem. Soc., 73, 5350 (1951); (b) L.G.S. Brooker, A. C. Graig, D. W. Heseltine, P. W. Jenkins, and L. L. Lincoln, ibid., 87, 2443 (1965).

2. E. M. Kosower, ibid., 80 3253 (1958).

3. E. G. McRae, J. Phys. Chem., 61, 562 (1957).

4. R. L. Reeves and R. S. Kaiser, J. Org. Chem., 35, 3670 (1970).

5. R. L. Reeves and L.K.J. Tong, J. Amer. Chem. Soc., 84, 2050 (1962).

6. R. L. Reeves and L.K.J. Tong, J. Org. Chem., 30, 237 (1965).

7. F. Franks and D.J.G. Ives, Quart. Rev., 20, 1 (1966).

8. F. Franks, in "Physico-chemical Processes in Mixed Aqueous Solvents," F. Franks, Ed., American Elsevier Publishing Co., New York, 1967, p. 50.

9. E. M. Arnett, ibid., p. 105.

10. G. S. Hammond, J. Amer. Chem. Soc., 77, 334 (1955).

A MODEL FOR THE BINDING OF CHAIN-LIKE MOLECULES TO COMPACT

MACROMOLECULAR SURFACES

Nora Laiken(1) and George Némethy

The Rockefeller University

New York, New York   10021

Several theories have been developed to describe the interactions between flexible polymer chains and surfaces(2-4). However,
these theories generally use uniform and essentially infinite planar
lattices to represent the surface, a restriction which prevents
their application to situations in which the binding surface either
is limited in size or covers the surface of a compact particle.
Examples of the latter case include colloidal particles and proteins,
many of which may be regarded as spheres or ellipsoids. It is the
purpose of this paper to present a model for the binding of chain-
like molecules to the surfaces of small, compact particles.

Actually, the earlier theories (2-4) contain additional features
which could limit their application to many systems, especially the
interactions between chain-like molecules and proteins. First of
all, the surface sites commonly, although not always(2), are regarded
as equivalent in their ability to adsorb polymer segments, and the
polymer chain is said to be composed of identical segments. In
contrast, many macromolecular surfaces have several different types
of binding sites (for example, a protein surface will contain functional groups from a variety of amino acids) and many flexible polymers contain segments which differ chemically (for example, the substituted alkanes consist of ionic or polar groups attached to a
hydrocarbon chain). A second limitation of most theories is that
they are formulated only for the limiting case of infinite or very
long chains. Many polymers employed in binding studies are relatively short (for example, the substituted alkanes used in protein
binding studies usually contain fewer than 20 carbon atoms).
Finally, although there are exceptions(2), most models emphasize
the properties of a single polymer chain at a surface, and the
binding of additional molecules, if treated at all, is assumed to

occur in the absence of competition for surface sites among adjacent molecules. In contrast, when the binding surface is small, the independence of successively bound molecules cannot be assumed automatically, and the effects of occupied sites on the configurations available to polymers bound at high levels of saturation can be appreciable.

The model to be presented here for the interactions between polymer chains and the surfaces of compact particles has certain formal similarities to the earlier theories(2-4) of flexible polymer binding (for example, the use of a lattice to enumerate polymer configurations). However, it includes several features which make it especially suitable for dealing with the above-mentioned interactions. In particular, it (i) is adapted to the treatment of finite, closed surfaces and short polymers, (ii) permits the introduction of chemical heterogeneity in the polymer segments and the surface, and (iii) allows for the dependence of a polymer's configuration on those of its previously bound neighbors. A description of the features of the model will be presented first, followed by a discussion of some of the mathematical details. Finally, some applications of the model will be described. Although the model originally was designed as a theory for the interactions between proteins and substituted alkanes(5-8), its potential applicability to a wide variety of systems should be evident.

DESCRIPTION OF THE MODEL

In order to place a planar lattice upon a closed surface, it is necessary to introduce a small number of irregularities into the lattice pattern. Variations in either the coordination number of certain sites or the spacing between sites, or both, are possible. As an example of an alteration in coordination number, a square lattice may be placed upon a cube by introducing tri-coordinated sites at the eight corners (Figure 1a). The placement of a square lattice on a globe to demarcate longitude and latitude provides an example in which irregularities have been introduced in both the spacing between adjacent lattice sites and the coordination number of the (two polar) sites (Figure 1b).

In choosing a lattice to represent the surface of a spherical particle, irregularities in coordination number (cf. Figure 1a) are to be preferred to large irregularities in the spacing between lattice sites (cf. Figure 1b), since the bond lengths between adjacent segments of most flexible polymers are constant. Therefore, the square lattice illustrated in Figure 1b is not suitable for the surface of a spherical particle. The lattice chosen for the present model was obtained by introducing irregularities in the coordination number of a small number of sites on a planar, triangular (hexacoordinated) lattice, allowing its placement on a sphere as illus-

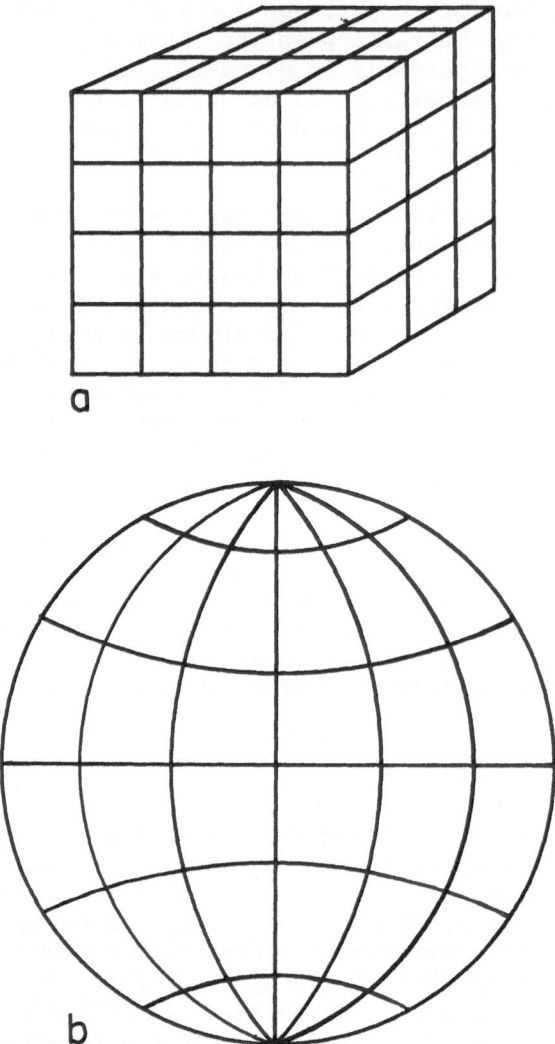

a

b

Figure 1. The placement of planar lattices on closed surfaces.
a) A square lattice may be placed upon a cube by introducing tri-
coordinated sites at the eight corners. b) A square lattice may be
placed upon a sphere by altering the spacing between adjacent lattice
sites and changing the coordination number of the two polar sites.

trated in Figure 2 (9). It should be mentioned that with the proper
definition of the statistical polymer segment(10), a triangular
lattice becomes suitable for the geometry of most flexible polymers(6).

The lattice of Figure 2 actually represents the outward projection of the sites on an icosahedron(9), the regular polyhedron with

Figure 2. The placement of a planar, triangular (hexa-coordinated) lattice upon a sphere by introducing a small number of irregularities in coordination number (penta-coordinated sites) at regular intervals (reproduced from Vol. XXVII; Cold Spring Harbor Symposia, 1962, by permission).

20 equilateral triangular faces and 12 vertices (Figure 3). If each face is subdivided into D^2 equilateral triangles (producing a division of each edge into D segments), a lattice is formed which contains $10(D^2-1)$ hexa-coordinated sites and 12 penta-coordinated sites, the latter located at the vertices.

In addition to their role in adjusting a planar, triangular lattice to a spherical surface, the penta-coordinated sites serve as reference points on the surface, dividing it into triangles. Within each triangle, each hexa-coordinated site may be classified according to its positional relationship to the penta-coordinated sites. This classification allows various arrangements of the different types of chemical groupings found on the macromolecular surface to be studied, with each such grouping represented by one or more classes of sites.

In many cases, it would be unrealistic to say that the entire surface of the particle is capable of binding polymers. For example, in the interactions between substituted alkanes and proteins, there is evidence(11-12) that the hydrocarbon segments of the polymer

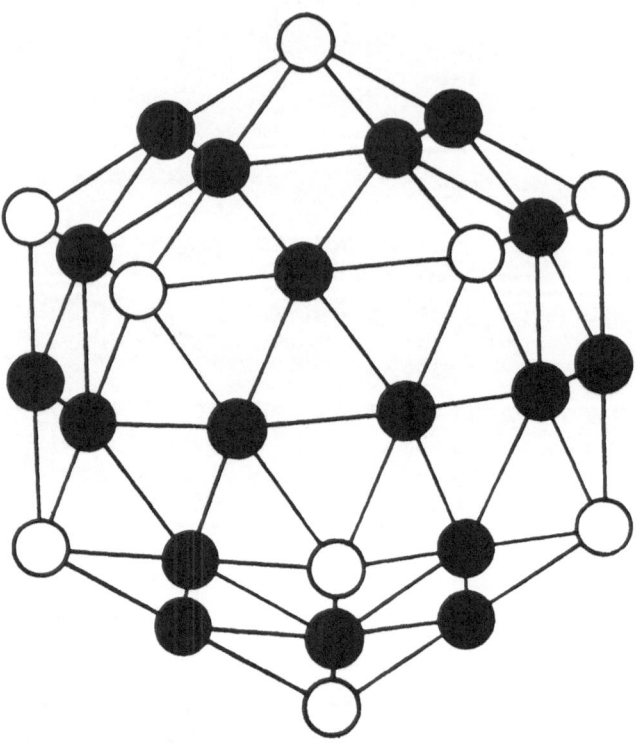

Figure 3. An icosahedron, the regular polyhedron with 20 equilateral triangular faces and 12 vertices (marked in white). Each triangular face of this icosahedron has been subdivided into four
equilateral triangles ($D^2=4$), thereby creating additional sites on
the lattice (marked in black). All sites are hexa-coordinated except for the irregularities (penta-coordinated) at the vertices.

bind primarily to nonpolar regions of the protein surface (composed
of side chains from the nonpolar amino acids). In general, only a
small percentage of the protein surface will be composed of these
amino acids due to their hydrophobicity(13). To treat cases such
as these, in which only a restricted area of the macromolecular
surface is capable of binding polymer segments, it is possible to
project a single triangular face of the icosahedron, or part of it,
onto a sphere to give a patch of sites, as illustrated in Figure 4.
The projection of several nonadjacent patches also is possible,
allowing a comparison of the binding data expected with and without
interference between adjacent polymer molecules. It is to be noted
that the model is not restricted to spherical particles when one or
more isolated patches of sites are used, since they could be placed
on a surface of any shape. Furthermore, a single patch of sites
could be used to represent a small planar molecule, such as a rigid
aromatic dye which binds chain-like molecules.

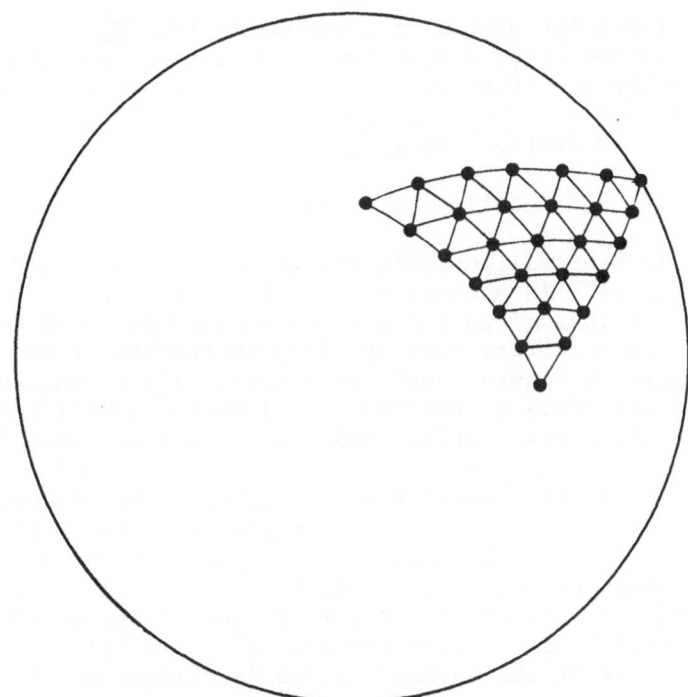

Figure 4. The projection of a single triangular face of an icosahedron onto a sphere, to give a patch of sites.

THEORETICAL TREATMENT

The grand canonical partition function for the system, which
consists of one particle in equilibrium with a solution of polymer
chains of fixed chemical potential, may be written in the form

$$\Xi = \sum_{N=0}^{M} \frac{1}{N!} \left(\prod_{i=0}^{N} q(i) \right) e^{\mu N/kT} \tag{1}$$

where M is the maximum number of polymers which can bind to the
macromolecular particle (determined by the size of the surface lattice and the number of segments in the polymer), μ is the chemical
potential of the polymer (which is directly related to its concentration in solution), and $q(i)$ is the canonical partition function
for the ith bound polymer. To evaluate the latter, the configurations of the polymer chains are classified according to their distribution of adsorbed and desorbed segments, which is referred to as
the state of adsorption (3) of the chain, S

$$S = \{d_0 , s_1 , d_1 , \ldots\ldots, s_{\ell-1} , d_{\ell-1} , s_\ell , d_0'\} \tag{2}$$

where s_i and d_i are the number of segments in the <u>ith</u> adsorbed
stretch and desorbed loop, respectively, d_0 and d_0' are the number
of segments in the two free ends of the polymer, and ℓ is the number
of adsorbed stretches. A statistical weight, $t_i(S)$, is assigned
to each state of adsorption, so that

$$q(i) = \sum_{\{S\}} t_i(S) \tag{3}$$

The summation is over all possible states of adsorption for a given
polymer chain. $t_i(S)$ is a count of all configurations available to
the <u>ith</u> polymer with a given state of adsorption and also contains
Boltzmann weighting factors corresponding to the number and kind of
surface site-polymer segment contacts formed. The dependence on i
arises because the previous binding of i-1 polymers will affect the
number of possible configurations and vacant lattice sites.

The terms $t_i(S)$ are calculated by regarding the placement of
each polymer on the lattice as a Markov process, each step of
which corresponds to the placement of a polymer segment on a lattice
site. This allows the use of matrices to enumerate the polymer con-
figurations(8). The states of the Markov process are defined in
terms of the lattice site classification discussed briefly above;
thus, the <u>ith</u> state of the Markov process is defined as the placement of a polymer segment on a class i site. The exact elements
used in the matrices are determined by the size of the lattice; in
addition, they differ somewhat for a single patch of sites (with
edges) and for a lattice which covers the entire surface (no edges).

From the partition function (eq 1) it is possible to calculate
the average number of polymers bound per macromolecular particle,
$\bar{N}$, as a function of c, the concentration of free polymer (corresponding to an experimental binding isotherm)

$$\bar{N} = kT \frac{\partial \ln \Xi}{\partial \mu} \tag{4}$$

equilibrium constants for the binding of successive polymers, and
parameters describing the average configuration of the bound polymers, such as average fraction of segments adsorbed to the surface,
average number and length of loops, and average length of desorbed
ends(8).

RESULTS AND DISCUSSION

Although the model presented here allows the treatment of heterogeneous systems, in which the sites differ chemically and the polymer chains contain one or more types of segments, it is illustrative
to begin with a discussion of homogeneous systems, in which all site-
segment interactions are equivalent.

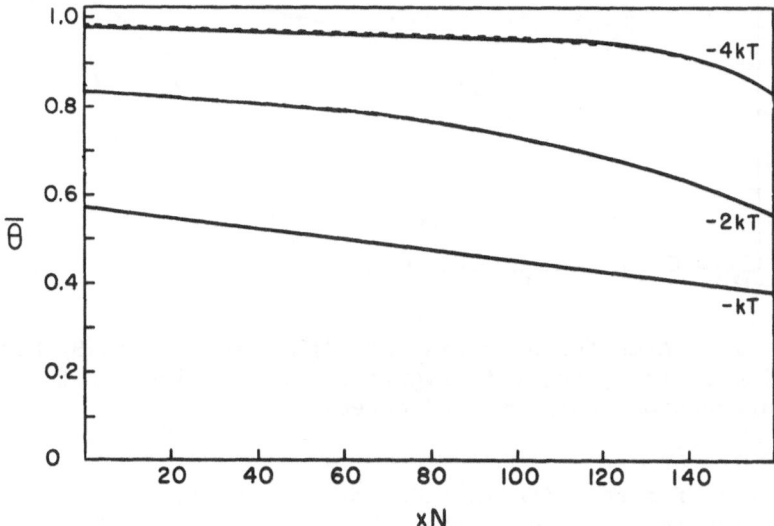

Figure 5. The variation in $\bar{\theta}$, the average fraction of polymer segments adsorbed, as a function of the number of segments belonging
to bound chains, xN , where x is the number of segments in the polymer. All curves were obtained with a polymer of eight segments
(x=8) binding to a lattice of 162 sites. The value of the site-
segment interaction energy is indicated by each curve.

An important factor in determining the equilibrium behavior is
the variation in the average configuration of the polymers as additional chains are bound to the surface. This variation is illustrated in Figure 5, in which $\bar{\theta}$, the average fraction of polymer
segments adsorbed, is plotted as a function of the number of segments belonging to bound chains. The curves represent the binding
of chains of eight segments to a surface with 162 identical lattice
sites for three different values of the site-segment interaction
energy. As N increases and the number of vacant surface sites decreases, bound segments experience a large reduction in their configurational entropy. A smaller amount of configurational entropy
is lost when groups of segments form loops above the surface since
more sites for the placement of segments are available there. Consequently, as N increases the chains prefer configurations with a
smaller fraction of their segments adsorbed, and $\bar{\theta}$ decreases. However, if the site-segment interaction energy is very negative, $\bar{\theta}$
remains fairly high in spite of the loss of entropy until near-sat-
uration is reached.

As a result of this variation in the average configuration of
the bound chains, the binding isotherms (expressed as $\bar{N}/M$ vs. log c)
are asymmetrical (Figure 6). The isotherms broaden as saturation
of the lattice is approached, since the binding of molecules at

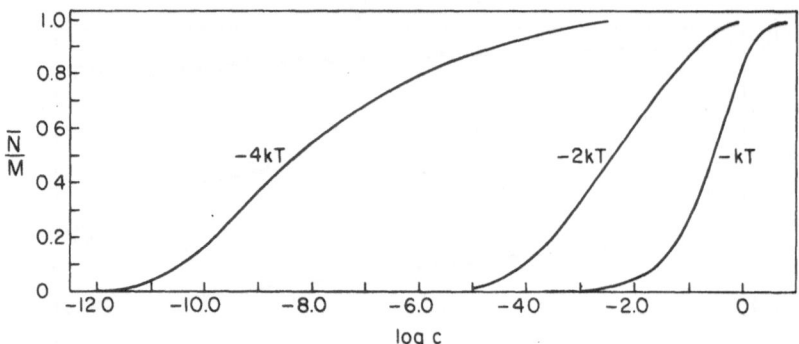

Figure 6. Isotherms for a polymer of eight segments (x=8) binding to a lattice of 162 sites (cf. Figure 5), with values of the site-segment interaction energy as indicated.

higher levels of saturation is made more difficult by the presence of the previously bound molecules. Because the $\bar{\theta}$ values are higher for more negative values of the site-segment interaction energy (Figure 5), the broadening is most marked at these values.

The ability to treat heterogeneous systems allows the application of the model to a wide variety of physical systems. Probably the most significant application and the one that has been investigated most thoroughly(5-8) is the binding of substituted alkanes (carboxylic acids and detergents) to protein surfaces. It is these interactions which will be discussed here. Thus, the chain consists of hydrocarbon segments and an ionic segment (carboxylate or sulfate), while the surface may include nonpolar, polar (noncharged), and ionic (charged) sites, representing the functional groups found on the amino acid side chains and the peptide backbone. Approximate values for the site-segment interaction energies are calculated as a sum of contributions from electrostatic interactions, nonbonded (van der Waals) interactions, and energies of desolvation(6). Because of the latter factor, the energies of interaction and the temperature dependence of binding are strongly dependent on the presence of an aqueous solvent. This is demonstrated in Table I, which illustrates the various contributions to ΔG^o , the free energy of binding of an alkyl (C_2H_4) group to a nonpolar surface. It is assumed that the free energy of the C_2H_4 group-nonpolar site interaction, $\Delta G^o_{H\phi}$=-1.3 kcal/mole, arises from hydrophobic interactions (14) in an aqueous solvent and thus is composed of both an energy and an entropy contribution. Actually, for the binding of a flexible chain the value of $\Delta G^o_{H\phi}$ is only -1.3θ kcal/mole of C_2H_4 groups since not all parts of the chain are in contact with the surface. Due to the large entropy contribution to the hydrophobic interaction, the net entropy of binding is positive in spite of the loss of configurational entropy upon binding. Since the enthalpy also is positive, this corresponds to a strengthening of binding with increasing

TABLE I

Approximate Changes in the Thermodynamic Parameters
(per C_2H_4 group[a]) for the Binding of an Alkyl Chain Molecule
to a Nonpolar Surface of a Compact Particle, at 25° C

Parameter			Binding in Aqueous Solution	Binding in Nonaqueous Solution
$\bar{\theta}$			0.83	0.83
ΔG^o	kcal/mole		-0.56	-0.56
$\Delta G^o_{H\phi}$	b	"	-1.13	--
ΔE_{vdw}	b	"	--	-1.13[c]
ΔG_{conf}		"	0.57	0.57
ΔS_{conf}	cal/deg·mole		-1.9	-1.9
$\Delta S_{H\phi}$		"	7.6	--
ΔS^o_{total}		"	5.7	-1.9

[a] Used as the statistical segment(10) in the calculations(6).

[b] Corresponds to the site–segment interaction energy of Figures
5 and 6. The value used here is very close to -2kT at 25° C.

[c] Assumed here for the sake of comparison with the data in aqueous
solution.

temperature (near 25° C). Table I also includes the corresponding
data for binding in the presence of a nonaqueous solvent. For the
sake of comparison, it was assumed that the free energy of the
C_2H_4 group-nonpolar site interaction is the same, i.e. -1.3 kcal/mole,
but that it arises entirely from van der Waals interactions (ΔE_{vdw}).
In this case, the only contribution to the entropy is configurational
and the temperature dependence is the opposite of that observed when
an aqueous solution is present.

The isotherm for the binding of hexadecanoic acid to a surface
composed of a large number of nonpolar sites and 12 positively charged
ionic sites ($-NH_3^+$) is shown in Figure 7. The curve is biphasic,
since an intermediate plateau occurs when $\bar{N}=12$, the number of ionic
sites, followed by another rise to the final plateau when the entire
lattice is saturated (when 18 chains are bound). The first phase
of the curve represents the binding of chains with carboxyl segments
bound to the charged sites, while in the second phase the chains
bind with their carboxyl segments desorbed (due to the prior saturation of the charged sites). This interpretation is verified in a
study of $\bar{\theta}$ and other configurational parameters for this system(6).

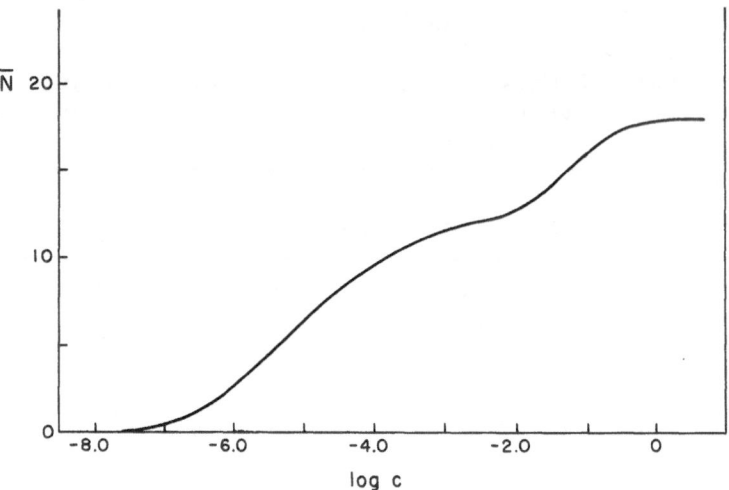

Figure 7. Isotherm for hexadecanoic acid binding to a lattice with 12 positively charged ionic sites and 150 nonpolar sites.

Higher values of c are required for binding in the second phase, since a smaller energy gain is possible per chain without the very favorable charge-charge interaction.

If polar sites are introduced on the surface (representing the carbonyl and amide groups of the peptide backbone), the first phase of the curve shifts to higher values of c, since the number of non-polar sites is reduced by introducing polar sites and a smaller amount of energy is gained by those chains which bind with some of their hydrocarbon segments attached to polar sites (hydrocarbon segment-polar site interactions are less favorable than hydrocarbon segment-nonpolar site interactions). On the other hand, the presence of polar sites makes binding in the second phase more favorable and therefore shifts it to lower c values, because of the energy gained from the carboxyl segment-polar site interactions (without the polar sites, the carboxyl segment would be unbound in the second phase).

If a sufficiently large number of polar sites is introduced (for example, if less than one-half of the lattice sites are non-polar), the intermediate plateau disappears almost entirely. In other words, the presence of an intermediate plateau in the binding curves is dependent upon the existence of a large number of the sites of lowest energy relative to the number of high energy sites, so that the former are only partially saturated when the latter are saturated completely.

Experimentally, biphasic isotherms have been observed for the binding of carboxylic acids to proteins(15), although the complete second phase never has been demonstrated due to the limited solubility of these chain molecules. These experiments, and others with similar systems(16-20), have been analyzed according to the commonly used theory of multiple equilibria(21-22). The latter theory is formally equivalent to the Langmuir theory of the ideal lattice gas, since the internal degrees of freedom of the bound molecules are neglected and it is assumed that the molecules bind independently. Thus, the binding isotherm is given by the equation

$$\bar{N} = \frac{Mkc}{1+kc} \tag{5}$$

when all of the binding sites are identical, or

$$\bar{N} = \sum_{i=1}^{n} \frac{M_i k_i c}{1+k_i c} \tag{6}$$

when there are n classes of binding sites. $\bar{N}$ and c are defined as above and M_i is the maximum number of chains bound, with intrinsic association constant k_i, in the ith class of sites. Eq 6 is used to analyze binding data such as those described above(15-20).

However, in many cases it is necessary to take into account the configurational flexibility of the bound molecules, so that the model presented here has to be used to describe experimental data instead of eqs 5 and 6. This is demonstrated in Figure 8, in which the change in free energy for the binding of the first chain to the surface is plotted as a function of the number of carbon atoms in the chain. The model discussed in this paper, which includes the ability of the chain molecules to bind in a large number of configurations, predicts an increase in the strength of binding with increasing chain length, as observed experimentally(15-20,23). On the other hand, if a calculation is carried out for rigid binding of chains in a single conformation (ref. 5, Appendix B), this being the equivalent of eq 5, the strength of binding decreases with increasing chain length. Cases in which eqs 5 and 6 may be applicable to the binding of flexible chain molecules have been discussed elsewhere(7). The trends indicated in Figure 8 are not affected by such details as the size of the lattice or the arrangement of the sites.

An increase in the strength of binding with increasing chain length also has been observed for the binding of chain molecules to planar aromatic dyes(24), indicating that the flexibility of the bound molecules is important in this case as well.

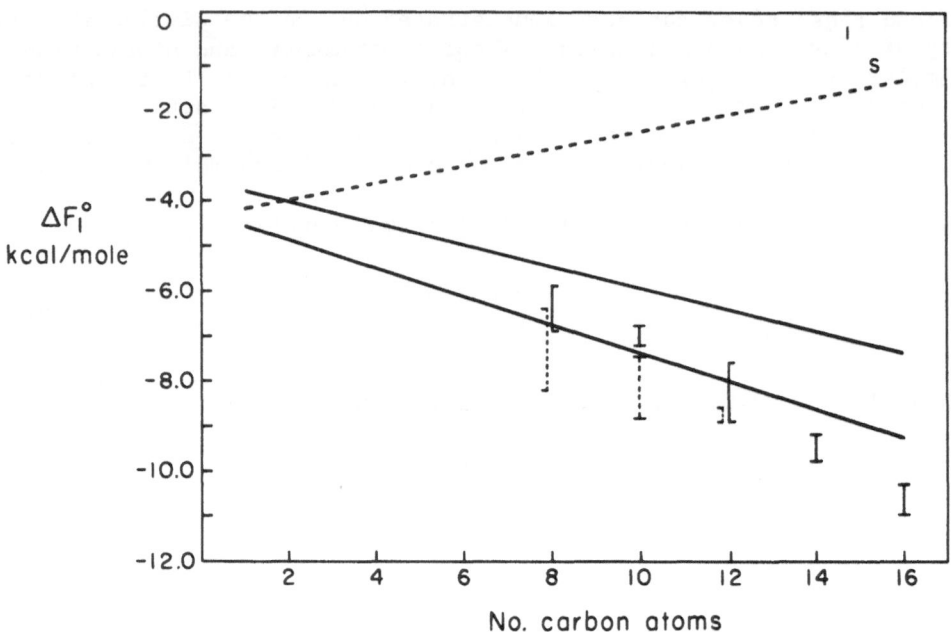

Figure 8. The variation in $\Delta F_1^{\,0}$, the change in free energy for the binding of the first chain molecule, with the number of carbon atoms in the chain. Experimental data for the binding of carboxylic acids (⊥) and alkyl sulfates (⫶) to serum albumin are shown, with the length of the vertical bars indicating the ranges of the experimental values reported in the literature(15-20,23)

The solid lines in Figure 8 are calculated using the present model for flexible binding; the upper solid line was obtained with a lattice of three positively charged ionic sites and 22 nonpolar sites, while the lower solid line represents binding to a lattice identical with that used in Figure 7 (12 positively charged ionic sites and 150 nonpolar sites). The dashed line is calculated assuming rigid binding of the chain molecules in a fixed conformation to the lattice of Figure 7. It is seen that the flexibility of the bound chains must be considered in order to account for the trend of the experimental data.

SUMMARY

A statistical-mechanical model has been developed to describe the binding of chain-like molecules (ligands) to macromolecular surfaces. To represent such surfaces, a planar lattice is folded over a sphere by introducing a small number of irregularities. This allows the treatment of compact macromolecules since the surface is not infinite as was always assumed previously. The model also permits the consideration of a very small array of lattice sites. Each lattice site binds a single functional group or segment of the ligand. It is necessary to account for the ability of chain-like molecules to bind in a large number of con-figurations, including many in which not all segments are in contact with the surface (e.g. some of the segments may form desorbed loops). The configurations are enumerated by counting the possible arrangements on the lattice (Markov process). The relationship between the concentration of free ligands in solution and the average number of ligands bound to each macromolecule (corresponding to an experimental binding isotherm) may be calcu-lated, as well as parameters which characterize the average configuration of each bound ligand (e.g. average fraction of segments adsorbed, average number of loops). Several features of the model make it especially suitable for analyzing interactions between proteins and substituted alkanes. The energies of inter-action and temperature coefficients for binding are a function of the presence of water. Some applications to such systems are discussed.

REFERENCES

1. Current address: Department of Chemistry, The University of Oregon, Eugene, Oregon 97403.
2. A. Silberberg, *J. Phys. Chem. 66* , 1872, 1884 (1962).
3. R.-J. Roe, *Proc. Nat. Acad. Sci. U.S. 53* , 50 (1965); *J. Chem. Phys. 43* , 1591 (1965).

4. Further references are included in references 2 and 3.

5. N. Laiken and G. Némethy, *J. Phys. Chem. 74 *, 4421 (1970).

6. N. Laiken and G. Némethy, *J. Phys. Chem. 74 *, 4431 (1970).

7. N. Laiken and G. Némethy, *Biochemistry* , in press.

8. For details not discussed here, references 5-7 should be consulted.

9. D.L.D. Caspar and A. Klug, *Cold Spring Harbor Symp. Quant. Biol. 27 *, 1 (1962).

10. P.J. Flory, "Principles of Polymer Chemistry," Cornell University Press, Ithaca, New York, 1953.

11. R.M. Rosenberg, H.L. Crespi, and J.J. Katz, *Biochim. Biophys. Acta 175 *, 31 (1969).

12. G. Gillberg-LaForce and S. Forsén, *Bioch. Biophys. Res. Comm. 38 *, 137 (1970).

13. W. Kauzmann, *Adv. Prot. Chem. 14 *, 1 (1959).

14. G. Némethy, *Angew. Chem. (Int. Ed.) 6 *, 195 (1967).

15. D.S. Goodman, *J. Amer. Chem. Soc. 80 *, 3892 (1958).

16. J.D. Teresi and J.M. Luck, *J. Biol. Chem. 194 *, 823 (1952).

17. A. Ray, J. Reynolds, H. Polet, and J. Steinhardt, *Biochemistry 5 *, 2606 (1966).

18. J. Reynolds, S. Herbert, H. Polet, and J. Steinhardt, *Biochemistry 6 *, 937 (1967).

19. J. Reynolds, S. Herbert, and J. Steinhardt, *Biochemistry 7 *, 1357 (1968).

20. A. Spector, K. John, and J. Fletcher, *J. Lipid Res. 10 *, 56 (1969).

21. I.M. Klotz, *in* "The Proteins," Vol. IB, H. Neurath and K. Bailey, Ed., Academic Press, New York, New York 1953.

22. J. Steinhardt and J. Reynolds, "Multiple Equilibria in Proteins," Academic Press, New York, New York, 1969.

23. F. Karush and M. Sonenberg, *J. Amer. Chem. Soc. 71 *, 1369 (1949).

24. R.L. Reeves, personal communication.

EFFECT OF STATE OF CONDENSATION ON THE IONIZATION POTENTIAL OF ASSOCIATED LIQUIDS

M. Anbar[*], G. A. St. John[*], H. R. Gloria[**] R. F. Reinisch[**]

*Stanford Research Institute

**NASA-Ames Research Center

ABSTRACT

Liquid water and alcohols have been shown to undergo photo-ionization when subjected to vacuum ultraviolet in the range 210–180 nm. Hydrated electrons have been identified as products of photolysis in liquid water by a number of characteristic reactions using competition kinetics as a quantitative criterion. The reactions $e_{aq}^- + SF_6$, $e_{aq}^- + H_3O^+$, and $e_{aq}^- + Cd^{++}$ have been used as the principal analytical tools of investigation. The photoproduction of electrons was shown to be the result of a single photon process. A parallel investigation of the photodissociation of water to H + OH suggests that photoionization and photodissociation of liquid water involve the same excited state; the cross section for photodissociation being about ten times larger than for photo-ionization.

Photoionization of liquid water occurs at energies significantly (> 6.5 eV) below the ionization potential in the gas phase. The photolysis of D_2O and of methanol shows analogous behavior. It seems that the association in the liquid state lowers the ionization potential mainly by solvation of the positive ion, H_2O^+, though the formation of excimers should also be considered.

INTRODUCTION

The first absorption continuum of liquid water in the far ultraviolet peaks at about 165 nm and tapers off down to about 200 nm.[1-3] The absorption of photons in this range results in a substantial photodissociation of liquid water to H + OH ($\varphi \sim 0.5$).[4,5]

Using flash photolysis it has been shown that photoionization also takes place in the same spectral range.[6]   In their experiments Boyle et al.[6] were able to demonstrate the transient absorption spectrum of hydrated electrons, e_{aq}^-, formed as a direct result of photoabsorption.   In fact, Boyle et al. were not the first to observe the photoionization of liquid water.   Hydrated electrons were observed by Matheson et al.[7] in the flash photolysis of methanol-containing solutions (0.2 M).   The very small yields obtained and the presence of methanol prevented at the time a positive identification of water as the source of e_{aq}^-.   In view of our results described in this paper, it is very likely that the e_{aq}^- observed did originate from the photolysis of water.   In two steady state studies on the photolysis of water there was an indication of the formation of e_{aq}^-.[5,8]   Sokolov and Stein[5] set an upper limit of $\varphi(e_{aq}^-) < 0.05$, but they were limited by the absorption of N_2O--their e_{aq}^- scavenger-- which produces N_2 on photolysis, and by the relatively inadequate analytical procedures of determining small amounts of N_2 in the presence of much larger quantities of hydrogen.   Getoff and Schenk estimated $0.02 < \varphi(e_{aq}^-) < 0.04$ at 184.5 nm from a complex system of scavengers including CO_2 as e_{aq}^- scavenger.[8]   Because of a gross error in assessment of the relative absorption coefficients of H_2O and CO_2 (they used ε_{H_2O} for water vapor which is about 3 orders of magnitude higher than that of liquid water), their measured quantum yields are expected to be low.

Although the quantum yield of e_{aq}^- in the photolysis of water above 180 nm remained an open question, ranging from 0.004[6] to 0.045,[5] there were strong indications that water undergoes photo-ionization, although with a very small yield, some 6 eV below its ionization potential in the gas phase.   Such a rather unexpected result required independent verification that the electrons observed are not due to the photoionization of a trace impurity.   Even if found true, this observation poses a number of interesting questions as to the mechanism of the photoionization of water.   Is this a single photon process, or is it due to a photionization of an electronically excited species? Does this photoionization result from the same excited state which leads to photodissociation? Is the observed photoionization a manifestation of a far more extensive photoionization which results in H atom formation by a fast vicinal recombination of $H_2O^+ + e_{aq}^-$? Is this photoionization at such a low energy a unique phenomenon of water, or is it the property of other associated liquids as well?

Using a more sensitive as well as specific analytical technique than previously employed, it was possible in the present study to establish the photoionization of water above 180 nm beyond any reasonable doubt, to determine its quantum yield, and to get far better insight into the mechanism of photoionization of water. This process seems to be the manifestation of a more general phenomenon shared by other associated liquids.

EXPERIMENTAL

Light Source. The light source used was a 10-kw vortex stabilized high pressure argon plasma arc, "Plasmatron," with magnesium fluoride window (Gianini Scientific Corp., Santa Ana, Calif.)

Monochromators. For F^- production, we used a McPherson vacuum monochromator of the 1/2 meter Seya-Namiyoka configuration with the grating blazed at 150 nm, using 2 mm slits.

H_2 production was carried out with a Diffraction Products Monochromator, 1/2 meter, with 6.6 nm bandpass and grating blazed for 300 nm, flushed with flowing N_2.

Cutoff Filters. Cutoff filters consisted of solutions of NaCl, KBr, and KI. Analytical grade reagents were dissolved in triply distilled conductivity water and interposed between lamp and sample in a 1-cm Beckman quartz sample cell.

Neutral Intensity Filters. Perforated black anodized aluminum screens supplied with the Cary Spectrophotometer were used to decrease light intensity by known amounts (50, 32, and 8%).

Irradiation Cell. The photolysis was carried out in a 10-cm pathlength quartz cell with sapphire windows (manufactured by Unified Science, Pasadena, Calif.) attached to a pyrex sample vessel for degassing by 3 freeze-thaw cycles and saturating with SF_6.

Reagents.
1. Water - Conductivity water, Chematics Research, Reseda, Calif.
2. D_2O - Biorad Labs., Richmond, Calif. Double redistilled in all-glass apparatus under N_2.
3. Methanol - J. T. Baker Spectrophotometric Grade.
4. SF_6 - Matheson.
5. $HClO_4$, NaOH, reagent grade - Mallinckrodt; $CdClO_4$, $YbClO_4$ - Electronic Space Products, Inc., Los Angeles, Calif.

The radiolysis experiments were carried out with a Co^{60} gamma source with a dose rate of 0.8 krad/min determined by the Fricke dosimeter.

Rationale of Analytical Procedure. As a specific e^-_{aq} scavenger, we chose SF_6. Because of its high electron affinity[9] this compound has been used as an electron scavenger in gas phase studies[10] as well as in organic systems.[11] In aqueous solution SF_6

reacts with e_{aq}^- at a diffusion controlled rate (k = 1.65 $\times$ 10^{10} M^{-1} sec^{-1})[12] to give six equivalents of fluoride ions for every e_{aq}^- scavenged.[12] The following sequence of reactions was suggested to explain the formation of the six fluoride ions.[12]

$$SF_6 + e_{aq}^- \rightarrow SF_6^- \rightarrow SF_5\cdot + F^-$$

$$SF_5\cdot + 2H_2O \rightarrow SF_4 + F^- + OH + H_3O^+$$

$$SF_4 + 9H_2O \rightarrow SO_3^{2-} + 6H_3O^+ + 4F^-$$

$$\begin{cases} SO_3^{2-} + 2OH \text{ (or } H_2O_2) \rightarrow SO_4^{2-} + H_2O \\ \text{or} \\ SO_3^{2-} + SF_5\cdot + 2H_2O \rightarrow SO_4^{2-} + SF_4 + F^- + H_3O^+ \end{cases}$$

Hence each e_{aq}^- gives 6F$^-$ and, by monitoring F$^-$, advantage can be taken of this chemical amplification.

The rate of reaction of SF$_6$ with H atoms is relatively slow. Assuming that no H atom scavenger was present in the system,[12] SF$_6$ + H did not compete at all with the H + H and H + OH reactions. Thus SF$_6$ + H must proceed with a rate slower than 10^2 M^{-1} sec^{-1}, if it proceeds at all. SF$_6$ is transparent to light down to $\lambda >$ 142 nm.[13,14] In spite of its low solubility in water, it seems to be ideally suited for competition kinetics of photolytic electrons in aqueous solutions.

EXPERIMENTAL PROCEDURE

A 25-ml sample of triply distilled water was degassed in the pyrex bulb by repeated freeze-pumping to $\sim$ 10^{-5} torr. SF$_6$ was then introduced into the cell at atmospheric pressure and the water saturated by vigorous shaking. Under these conditions the concentration of SF$_6$ in the water reaches 2.2 $\times$ 10^{-4} M.[15] The water was then poured into the sapphire cell and irradiated with the light from the Plasmatron arc lamp with a MgF$_2$ window at a distance of 15 cm (arc to sapphire window). The space between the lamp and cell was continuously swept with N$_2$ to maintain transparency in the vacuum ultraviolet.

Fluoride ion concentrations in the photolyzed solutions of SF$_6$ were determined by an Orion fluoride specific ion electrode (94-09) in conjunction with an Orion single-junction silver-silver chloride reference electrode (90-01); the electrode potentials were measured by a Beckman Model GS pH-meter. The electrodes were calibrated with standard sodium fluoride solutions. These calibrations gave excellent straight lines in the range 10^{-7}-10^{-3} M F$^-$, exhibiting a 59-mv change in potential for each tenfold change in fluoride ion activity. Values from 10^{-6} to 10^{-7} can be reliably measured if the electrode is preconditioned by soaking and standardizing in fresh solution and not exposed to F^- concentrations above 10^{-6} after conditioning.

The yield of H_2 produced in methanol-containing solutions was determined by gas chromatography using a 10-ft $\times$ 1/4-in Poropak Q column at $50^\circ C$, 15 ml/min nitrogen flow, with a thermoconductivity detector at 160 ma current on an F&M Model 720 gas chromatograph. Calibration of the system was carried out as follows:

To test recovery of H_2 from solution a series of seven 50-μl samples were radiolytically produced in a H_2O-KBr solution, assuming $G_{H_2} = 0.45$. The H_2 was freeze-thaw released from solution into a closed, evacuated system with a liquid-nitrogen cooled trap. The H_2 was collected by a Toepler pump and transferred by syringe to the gas chromatographic inlet. The photolytically produced samples were handled in the same manner. The detector was calibrated before or after each sample by the direct injection of H_2 into the chromatograph.

RESULTS AND DISCUSSION

1. The Production of Fluoride Ions from SF_6 in Photolyzed Water as a Function of Time

The production of fluoride ions from SF_6 in photolyzed water was studied as a function of time at neutral pH using polychromatic light. Owing to the absorption coefficient of liquid water, the practical cutoff wavelength was as high as 175 nm under the experimental conditions. Light of shorter wavelength is absorbed in the first few microns (at 180 nm 1 mm of water has an optical density of 2, and at 175 nm 90% of the light is absorbed in 50 microns of water).[2] The e^-_{aq} scavenger is thus depleted to zero within a few seconds in the layer adjacent to the sapphire window producing at the same time a thin layer of $SO_4^=$ ions which act as a cutoff filter. It has been found that the rate of F^- production was cut by 70% when a filter of 1-cm water was interposed between the light source and the photolysis cell. Using the absorption curve of water,[2] this means an "effective" wavelength of 187 nm. In other words, the photolyzed system behaved as if all the light was of 187 nm. Because some of the absorption took place at considerably longer wavelengths, a significant part of the photolysis took place at shorter wavelengths but probably not much shorter than say 182 nm. It may be inferred that the contribution of the strongly absorbing boundary layers was limited and that all the active photons were absorbed in the 10-cm cell.

The F^- production was found linear with time from 5 to 60 minutes (Fig. 1). At the given geometry and light intensity,

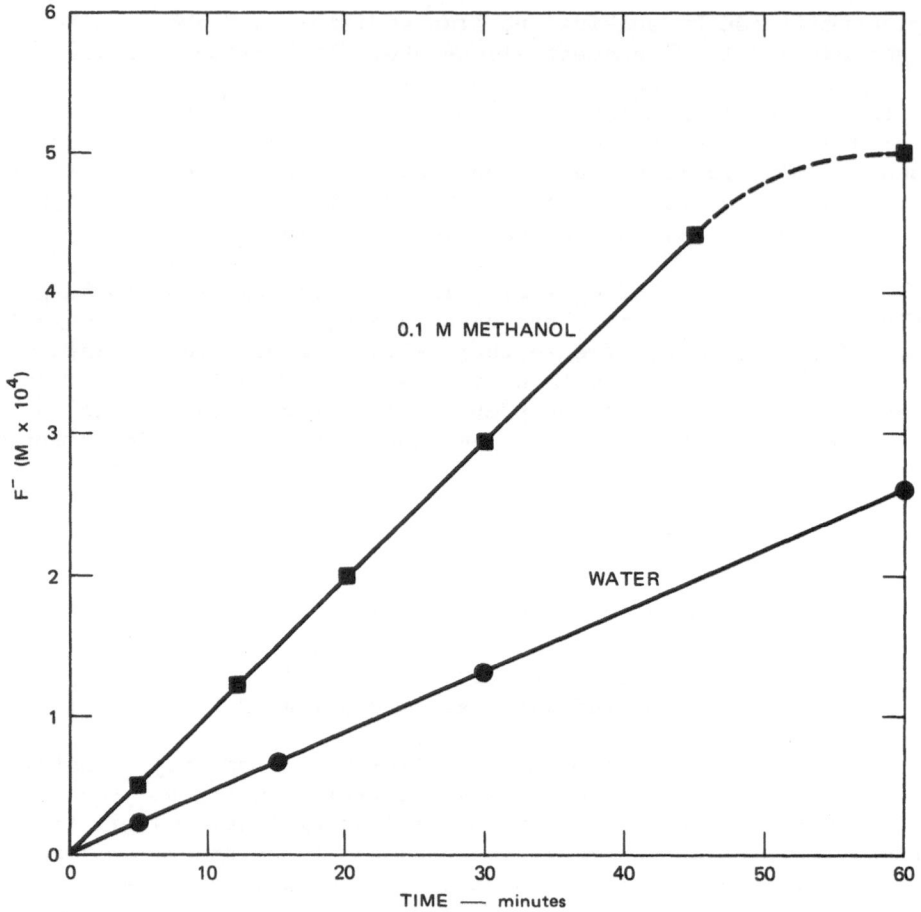

Fig. 1.   The Rate of Photolytic Fluoride Ion Production from SF_6
          as Function of Time in the Absence and Presence of 0.1 M
          Methanol

0.12 μmoles F^- (0.02 μmoles e_{aq}^-) were produced per minute. This
linear production of F^- indicates no exhaustion of reagent and no
buildup of interfering substances in one hour. At longer irradiation times the rate of e_{aq}^- production tended to decrease owing to
the buildup of H_3O^+. On the other hand, the formation of SO_4^{2-} added
a solute which generated e_{aq}^- on photolysis thus increasing the overall e_{aq}^- yield. By measuring the absorption spectrum of SO_4^{2-} in the
given spectral region, it was concluded that its contribution would
be minimal unless it reaches the 10^{-4} M level. As most of our work

with "polychromatic light" was carried out with irradiation times of 10 minutes, the final SO_4^{2-} concentration was lower than 10^{-5} M.

The same yield of fluoride ions, within 5%, was obtained for SF_6-saturated distilled water from three different sources: commercial water triple distilled; SRI Physical Sciences, triple distilled; and Ames, double distilled water. These results indicate that the hydrated electrons formed originate by photoionization of the water and not as a result of the photoionization of an adventitious impurity.

2. The Effect of Added Methanol or Ethanol on the Rate of F⁻ Production

The yield of F⁻ was found to increase when methanol was added to the water, and the F⁻ yield leveled off at [MeOH] = 0.03 M, staying constant up to 10^{-1} M MeOH (Fig. 2). A similar increase

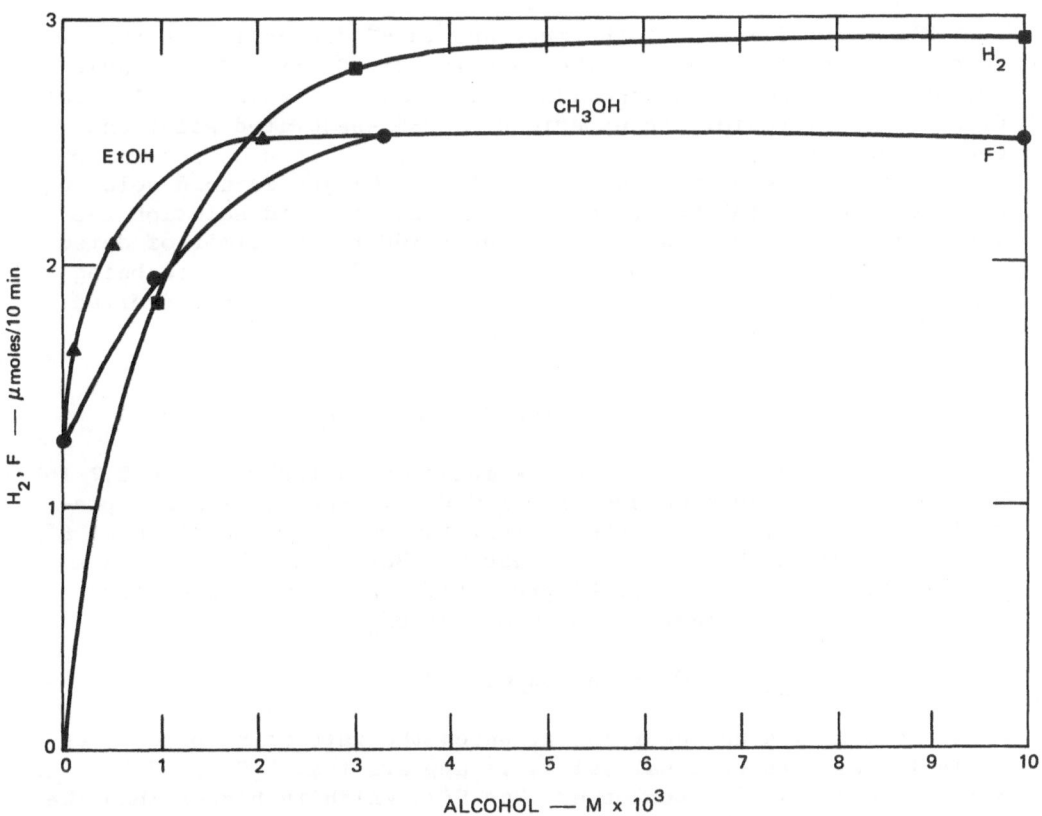

Fig. 2. The Rates of Photolytic Fluoride Ion and Molecular Hydrogen Production as Function of Added Methanol and Ethanol

was observed on addition of ethanol reaching the same limiting value of 0.25 μmole F^- (= 0.042 μmole e_{aq}^-) produced per minute. The effect of ethanol was, however, more pronounced than that of methanol; ethanol seemed to have an equivalent effect to methanol at about half the concentration. The most plausible explanation for the effect of these alcohols is that they scavenge OH radicals vicinal to the e_{aq}^-, thus inhibiting geminate recombination.

$$H_2O \xrightarrow{h\nu} H_2O^+ + e^- \rightarrow H_3O^+ + OH + e_{aq}^-$$

$$e_{aq}^- + OH \rightarrow OH^-$$

$$CH_3OH + OH \rightarrow H_2O + CH_2OH$$

This suggestion finds its support in the non-linear dependence of F^- production rate on alcohol concentration, which is typical for nonhomogeneous competition kinetics,[16] and by the twofold higher efficiency of EtOH, which is in line with the corresponding rates of reaction of OH radicals with MeOH and EtOH (k = 6.1 and 11.0 × 10^8 M^{-1} sec^{-1}, respectively[17]).

The possibility that the increase in F^- production in the presence of alcohols is due to the reaction of SF_6 with CH_2OH radicals, produced by hydrogen abstraction from methanol, has been ruled out by a series of radiolytic experiments. SF_6-saturated solutions containing 0.1 M MeOH were irradiated by gamma rays at neutral pH in the presence and absence of N_2O (1×10^{-2} M) and in acid solution (10^{-2} M $HClO_4$). In the presence of N_2O and in acid solution the yield of F^- dropped by a factor of over 100 to the limit of detection. Under the latter conditions, the CH_2OH radicals are being produced with a G value of about 6 but all the e_{aq}^- are converted into OH radicals or H atoms.

3.   The Effect of pH on the Rate of Production of F^-

The rate of F^- production was measured in the pH range 2.7 to 10.6 in absence and presence of 0.1 M MeOH. The results are presented in Fig. 3. It is evident that the rate of production of F^- is pH independent between pH 4.5 and 9. Below pH 4.5 there is a fast decline in the rate of F^- production, which may be easily explained by the competition of H_3O^+ for e_{aq}^-.

$$e_{aq}^- + H_3O^+ \rightarrow H + H_2O.$$

If we take pH = 4 as the point at which the rate of F^- production falls to half its original value, it appears that H_3O^+ is 2.2 times more efficient as e_{aq}^- scavenger than SF_6, which is higher than the value of 1.6 found by Asmus and Fendler[12] under radiolytic conditions. Now we have to remember that SF_6 at 2.2×10^{-4} M concentration picks up part of the e_{aq}^- from vicinal pairs with OH.

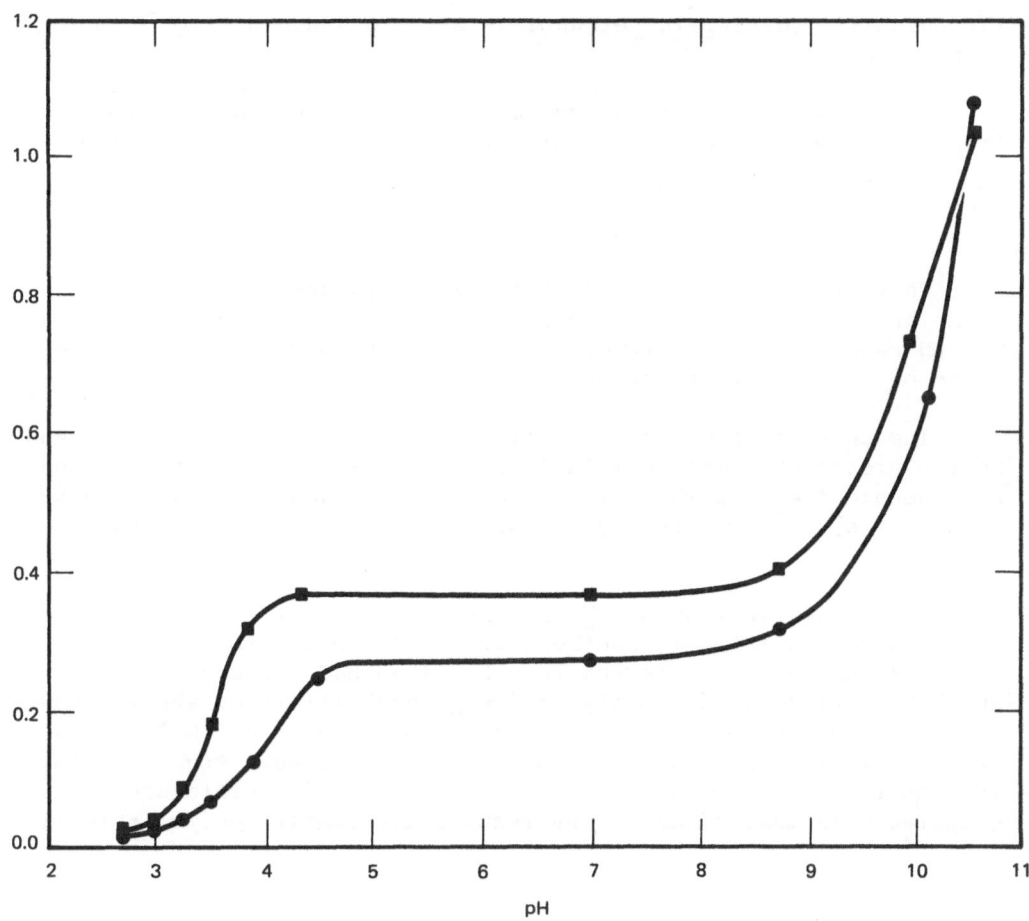

Fig. 3.   The Rate of Photolytic Fluoride Ion Production
as Function of pH

$$H_2O \xrightarrow{h\nu} e^- + H_2O^+ \rightarrow e^-_{aq} + OH + H_3O^+.$$

Under these conditions, the relative rates of diffusion of H_3O^+ and
SF_6 become critical and H_3O^+ becomes more competitive as its rate
of diffusion is higher by an order of magnitude. When methanol is
added, the vicinal OH radicals are scavenged and the relative effectiveness of H_3O^+ falls from 2.2 to 1.2, which is even lower than
1.4---the best ratio between the specific rate constants of e^-_{aq} with
H_3O^+ and SF_6 respectively.[18] This apparent higher scavenger efficiency of SF_6 (by about 15%) may be partly due to a correspondingly
slightly higher solubility of SF_6 in 0.1 M MeOH; we have found that

the solubility of SF_6 in methanol is about a hundred times higher than in water.

At the other end of the pH range the increase in F^- production (Fig. 3) is due to the photoionization of OH^- [7] as well as to the conversion of H to e_{aq}^- by the reaction[19]

$$OH^- + H \rightarrow e_{aq}^-.$$

This pathway for e_{aq}^- production is suppressed in the presence of methanol because 0.1 M MeOH effectively competes for H atoms in the pH range 9-11. The lower yields of F^- at high pH in the presence of MeOH are therefore expected.

The pH profile of the rate of F^- production in the absence and in the presence of methanol is thus completely consistent with the assignment of e_{aq}^- as the sole precursor of F^- and with the assumption that e_{aq}^- is produced by photoionization of H_2O in the pH range < 8.

The identification of e_{aq}^- as the precursor of F^- in our system has been corroborated by another series of competition kinetics experiments where e_{aq}^- scavengers other than H_3O^+ were used. Cd^{2+} and Yb^{3+}, which react rapidly with e_{aq}^-, have little UV absorption in the range 180-200 nm. Thus they could be added at 10^{-4} concentrations to the SF_6 photolyzed solutions. They were found to diminish the rate of F^- production and their scavenging efficiency was compared with that of SF_6. The results are summarized in Table 1.

Table 1

REACTION RATES OF e_{aq}^- AS CALCULATED FOR COMPETITION KINETICS[a]

	Pure Water	0.1 M MeOH	Literature Value[18]
SF_6	1.1 ± 0.1[b]	1.9 ± 0.2[b]	1.65
Cd^{++}	6.8 ± 2.2[c]	6.6 ± 2.0[c]	4.8-5.2
Yb^{3+}	–	3.9 ± 1.0[c]	3.7-4.3

[a]In units of 10^{10} M^{-1} sec^{-1}·

[b]Assuming $e_{aq}^- + H_3O^+ = 2.3 \times 10^{10}$ M^{-1} sec^{-1}.

[c]Assuming $e_{aq}^- + SF_6 = 1.65 \times 10^{10}$ M^{-1} sec^{-1}.

It is evident that, unlike the competition of H_3O^+, Cd^{++} ions do not show a significant change in apparent reactivity in the presence of 0.1 M MeOH. This corroborates our suggestion on the unique mode of action of H_3O^+ in scavenging e_{aq}^- from the vicinity of OH radicals. The slightly higher values of Cd^{++} compared with the pulse radiolysis data are in accord with other competition reactions of SF_6 where even the reactivity of H_3O^+ is found higher by over 10% from the pulse radiolysis value.[12] It is thus possible that the measured value of 1.65×10^{10} M^{-1} sec^{-1} for the $e_{aq}^- + SF_6$ reaction is slightly too high or that the solubility of SF_6 in water is somewhat lower than assumed.[15] In any case the agreement between the competition kinetics data and the absolute pulse radiolysis rate constants for H_3O^+, SF_6, Cd^{++}, and Yb^{3+} is sufficient to corroborate our conclusion that e_{aq}^- is the sole precursor of F^- in our photolyzed system.

4. The Formation of H_2 and the Quantum Yield of the Photoionization of Water

The rate of photodissociation of water was monitored by measuring the rate of H_2 evolution from photolyzed methanol-containing solutions. Under these conditions the following reactions take place:

$$H_2O \xrightarrow{h\nu} H + OH$$

$$CH_3OH + H \rightarrow CH_2OH + H_2$$

$$CH_3OH + OH \rightarrow CH_2OH + H_2O$$

$$2CH_2OH \rightarrow CH_3OH + HCOH \text{ or } (CH_2OH)_2.$$

The effect of methanol concentration on the H_2 yield is presented in Fig. 2. It can be seen that the H_2 yield levels off at $[MeOH] > 3 \times 10^{-2}$ M, just where the increase in F^- yield levels off. The production rate of H_2 is 1.16 times higher than that of F^- and therefore sevenfold higher than that of e_{aq}^-. Taking the quantum yield of H_2 in our spectral range in the presence of 0.1 M MeOH, $\varphi = 0.54$,[5] the quantum yield of e_{aq}^- is: $\varphi(e_{aq}^-) = 0.077$.

This quantum yield is substantially higher than that observed by Boyle et al.[6] using flash photolysis. A possible explanation is that in that investigation only the electrons which escaped from the "cages" were observed, whereas in our system, even in the absence of MeOH, SF_6 must have scavenged some electrons which would have otherwise recombined with their vicinal OH radicals or H_3O^+ ions. Interestingly enough, the relative yield of "free" $H/e_{aq}^- = 20$ obtained in flash photolysis is significantly higher than the ratio $H/e_{aq}^- = 7$ obtained by us in the presence of 0.1 M MeOH. The reason for the difference is most probably the greater chance for geminate recombination of e_{aq}^- with both H_3O^+ $(k = 2.3 \times 10^{10}$ M^{-1} $sec^{-1})^{18}$ and OH $(k = 3\times10^{10}$ M^{-1} $sec^{-1})^{18}$ compared with the single process H + OH $(k = 2\times10^{10}$ M^{-1} $sec^{-1}).^{21}$ In other words, e_{aq}^- "sees" in its cage two scavengers whereas the H atom encounters only one.

The quantum yield found by us in the presence of MeOH is higher than that estimated by Sokolov and Stein,[5] namely, $\varphi(e_{aq}^-) \leqslant 0.05$. The latter value was, however, obtained in a 1.3×10^{-3} M isopropanol solution where $\varphi(H_2)$ was just 0.33. Extrapolation of Sokolov and Stein's data for $\varphi(H_2) = 0.54$ would give $\varphi(e^-) \leqslant 0.08$. On the other hand, our quantum yield in the presence of 2.2×10^{-4} M SF_6 and in the absence of MeOH, $\varphi(e_{aq}^-) = 0.037$, is well below the upper limit of Sokolov and Stein. We have also to remember that their determination of $\varphi(e_{aq}^-)$ was carried out with a light source which emitted about 2×10^{-11} einstein sec^{-1} of the 184.9 nm photons into their photo-lyzing cell. Our light intensity was about 1.5×10^{-6} einstein sec^{-1}, in the same spectral range. This dramatic difference in light intensity allowed us to obtain substantial yields of e_{aq}^- where the previous study had to rely on marginal differences in the measured N_2 pressure. Sokolov and Stein should be complimented therefore for having obtained a result so close to ours with such inferior instrumentation and analytical technique. The other study[8] which has estimated $\varphi(e_{aq}^-)$ is much less reliable, as has been pointed out in the Introduction. The internal filtering effect of the scavenger, CO_2, might have resulted in low apparent quantum yield. Surprisingly, however, Getoff and Schenck's $\varphi(e_{aq}^-)$ estimate for 0.01 M formate + 0.02 M CO_2, $0.02 < \varphi(e_{aq}^-) < 0.04$ is still in fair agreement with our $\varphi(e_{aq}^-)$ measured in the absence of methanol.

5.   The Energy Threshold of Photoionization of Water

The yield of photolytic hydrogen and F^- in 0.1 M MeOH was measured as a function of wavelength.

The wavelength cutoff for the production of hydrogen was obtained with relatively little difficulty though the production was on the order of 0.1 μl H_2 in an hour compared to the polychromatic production rate of 50 μl in 10 minutes. Using 5-nm increments of wavelength advance, the production shows a peak at 190 nm. It then drops with wavelength to a cutoff at 203 nm (Fig. 4). Production of hydrogen also falls, as expected, below 190 nm because the high absorption of the water results in a pronounced self-filtering effect by thin layers which become depleted in SF_6.

Because of the detection limit of F^- $(\sim 10^{-7}$ M), the measurement of the wavelength cutoff of fluoride was carried out by the attachment of a McPherson Vacuum Spectrometer to the lamp in order to minimize the internally scattered light during the long exposures.

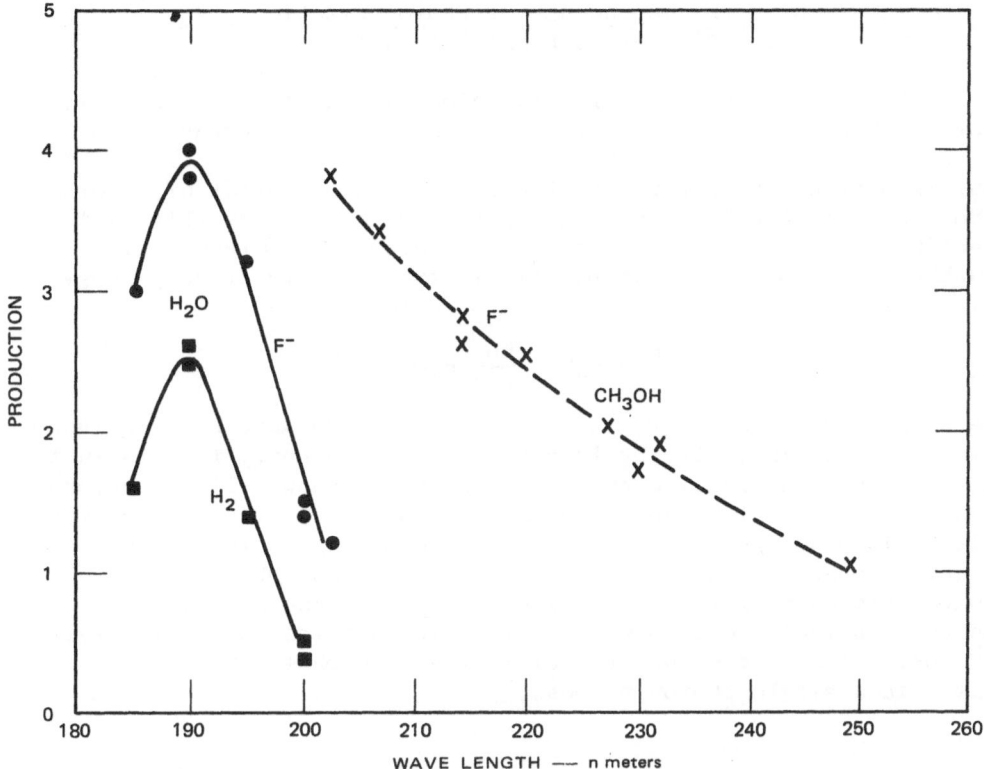

Fig. 4.   The Rates of Photolytic Fluoride Ion and Molecular
Hydrogen Production in H_2O and MeOH as Function of
Wavelength

By extrapolation we obtained a cutoff of F^- production at 206 nm
(Fig. 4).

Owing to the large experimental errors involved in the determination of H_2 and F^- close to the cutoff wavelength, the value of
the latter is estimated at 205±2 nm. In any case, it can be seen
from Fig. 4 that the photolytic yields of H_2 and F^- follow the same
wavelength dependence. It may be thus concluded that both photodissociation and photoionization of water originate from the same
excited state.

$$H_2O \xrightarrow{\ h\nu\ } H_2O^* \nearrow\quad H + OH$$
$$\searrow\quad H_2O^+ + e^-$$

6.   The Dependence of Photoionization of Water on Light Intensity

The cutoff wavelength for photoionization of water, 205 nm, corresponds to a quantum energy of 6.05 eV. This energy is in excess of the energy required to break the H-OH bond, 5.16 eV,[22] but is over 6.5 eV lower than the ionization potential of water in the gas phase (12.6 eV).[23] One simple explanation of this difference would be that photoionization of water is the result of a double photon process--the photoionization of a relatively long-lived electronically excited H_2O^* molecule.

$$H_2O \xrightarrow{h\nu} H_2O^* \xrightarrow{h\nu} H_2O^+ + e^-$$

We have studied therefore the yield of photoproduced F^- as a function of light intensity by inserting neutral density filters in the light path. The results presented in Fig. 5 show a linear dependence of the F^- yield obtained in a given time on the light intensity. In other words, the quantum yield of e_{aq}^- is independent of light intensity. This conclusion is further corroborated by the comparable quantum yields obtained by Sokolov and Stein,[5] who photolyzed their solutions with light intensities lower by a factor of 10^5. We must conclude therefore that the photoionization of water is a single photon process.

7.   The Photolysis of D_2O

A series of experiments was carried out in D_2O saturated with SF_6, both in the presence and absence of CD_3OD. D_2O absorption band is shifted by about 5 nm to the shorter wavelength region.[4,24]

The experiments with heavy water confirm the results with light water. Production of F^- in heavy water irradiated with polychromatic light is about 90% of that with light water. This 90% is the result of the lower light output of the lamp at the shorter wavelengths necessary to photoionize D_2O as well as the decreased transmission of the windows at these wavelengths. The deuterium, D_2, production closely follows the e_{aq}^- production in heavy water as in light water. The ratio of e_{aq}^- to $H_2(D_2)$ is 16±4% in D_2O compared with 14±2% in H_2O; the difference being insignificant.

Inserting a Beckman 1-cm quartz cell between the lamp and the sample cell decreases the rate of F^- production in D_2O by 40%, while reducing F^- yield in light water by 13%. The 1-cm cell filled with H_2O cuts the F^- in D_2O production to 12%, and 1 cm 10^{-3} M NaCl reduces it to 8%. The results are summarized and compared with the corresponding H_2O data in Table 2.

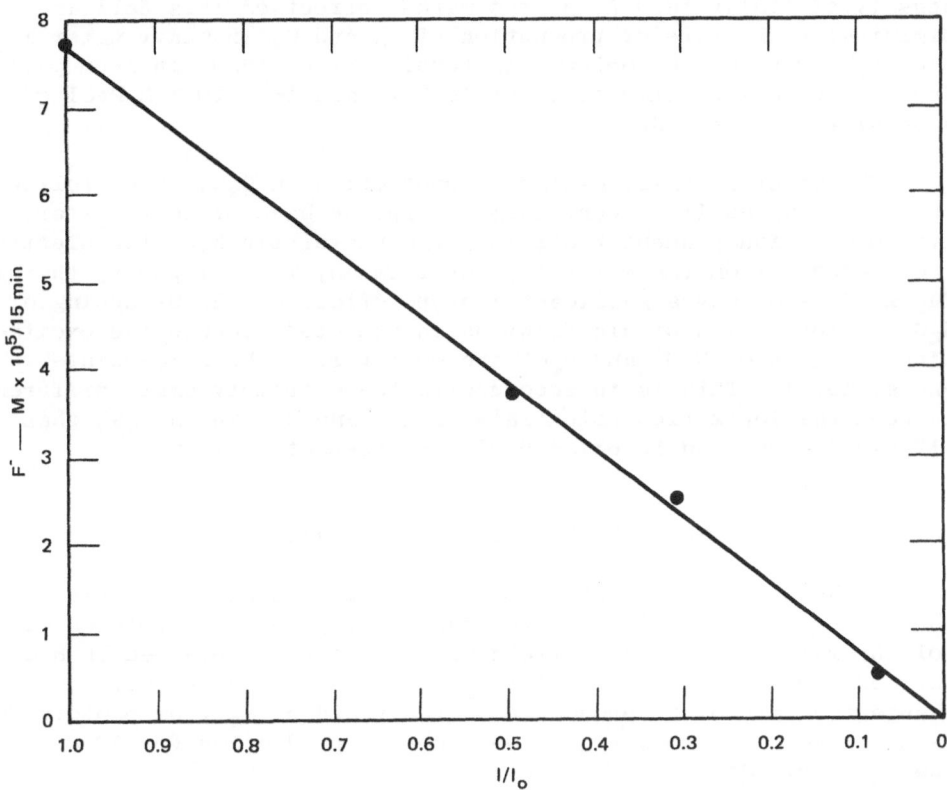

Fig. 5. The Effect of Light Intensity on the Rate of Fluoride in Production

Table 2

Filter	D_2O–CD_3OD				H_2O–CH_3OH			
	F^-		D_2		F^-		H_2	
Polychromatic, no filter	.58*	100%	.56*	100%	.66*	100%		100%
Quartz cell, empty	.35	60	.30	54	.57	87	.64	
1 cm H_2O	.07	12				34		29
1 cm 10^{-3} NaCl	.045	8				15		12

*Data in μmoles in 10 min radiation.

A Cary scan of the 1-cm quartz cell showed an absorption density of .13 at 1850 Å, so the marked effect of this cell in diminishing the rate of production of e_d^- and D_2 in heavy water is caused by wavelengths below this level. Water (H_2O) in 1-cm path-length showed a density of .7 at 1880 Å, and 1-cm 10^{-3} M NaCl gave a density of .7 at 1920 Å.

The spectral shift of the F^- production in D_2O, which follows the shift in the first continuum absorption band of heavy water, provides an independent proof that the photolytic hydrated electrons are produced from the water and not from any trace impurity therein. The absence of any significant isotope effect on the branching of H_2O^* to ionization or dissociation is expected because the excitation energies of H_2O^* and D_2O^* are so far from the zero-point-energy level. This is in accord with the extremely small difference between the ionization potentials of H_2O and D_2O in the gas phase (12.614 ± 0.005 and 12.637 ± 0.005 eV, respectively).[25]

8.   The Photolysis of Methanol

Methanol saturated with SF_6 was photolyzed under similar conditions to those of the water photolysis. When photolyzed with polychromatic light the F^- yield was 85% of that obtained in H_2O. The addition of 0.2, 1, and 10% H_2O to MeOH did not significantly change the rate of F^- production. The yield of H_2 from photolyzed methanol amounted to 5 times the fluoride production (or 30 times the e_{MeOH}^- yield).

In an independent series of radiolytic experiments, solutions of SF_6 in MeOH were irradiated in the presence and absence of N_2O and $HClO_4$. The fluoride yield was cut to less than 10% in the presence of these electron scavengers, indicating that, as in the water system, solvated electrons are the sole precursors of fluoride ions produced from SF_6.

The wavelength dependence of the photolytic F^- yield in methanol was determined and the results are presented in Fig. 4. The F^- formation seems to follow the first absorption continuum in methanol.[26]

In short, with the difference in the branching ratio of CH_3OH^* to give hydrogen or an electron, methanol seems to behave similarly to water and undergoes photoionization with a cutoff wavelength of approximately 265 nm (4.7 eV); this is again 6.1 eV lower than the ionization potential of methanol in the gas phase (10.83 eV).[27]

The photoionization of ethanol and other alcohols is presently under investigation.

9.   The Mechanism of Photoionization of Associated Liquids

It has been pointed out[6] that the combined energies of hydration of H_3O^+ and e^-_{aq} are sufficient to allow the photoionization of water with 6.05 eV photons, i.e., 6.55 eV below the ionization potential in gas phase.

$$2H_2O \rightarrow e^-_{aq} + H_3O^+_{aq} + OH_{aq} - 5.8 \text{ eV }[6]$$

Obviously this overall process is far too complicated to take place in a fast single step and it is certainly in violation of the Franck Condon principle. This mechanism involves a proton transfer reaction in addition to a series of rearrangements of water molecules around the newly formed H_3O^+ and e^-_{aq}. It requires, therefore, the formation of a long-lived, electronically excited H_2O^*, which managed somehow to orient its neighboring molecules to accept and solvate simultaneously a H^+ ion and an electron with just 0.25 eV excess energy. It is evident that this mechanism as it stands is highly improbable.

It is evident that the photoionization of liquid water is not an instantaneous process like that in the gas phase. In order to produce an electron plus a H_2O^+ ion 12.6 volts are required in the gas phase; and if we take into consideration the reduction in ionization energy owing to the higher dielectric constant of the polar medium, the ionization potential in the liquid phase may drop to > 9.5 eV. The latter estimate is based on the effect of the dielectric constant on ionization potentials[28] and on the differences, δ, of the ionization potentials of various aliphatic and aromatic hydrocarbons measured in the gas and in the liquid phases, which range $0.5 < \delta < 1.5$ eV.[28],[29] It has to be assumed, therefore, that the hydration energy of H_2O^+ and perhaps that of e^- make the photoionization feasible. The hydration energy of H_2O^+ can be estimated from that of H_3O^+:

$$\Delta H(H_3O^+_{aq}) - \Delta H(H_3O^+) = 11.3 \; [30] - 7.15 \; [31] = 4.15 \text{ eV}.$$

Because of its smaller radius, it is expected that the hydration energy of H_2O^+ will be higher than that of H_3O^+. Thus

$$\Delta H(H_2O^+_{aq}) - \Delta H(H_2O^+) > 4.15 \text{ eV}.$$

It is plausible that an electronically excited H_2O^* molecule is formed on photon absorption which then polarizes its adjacent H_2O molecules, making them ready to accept the newly formed H_2O^+. We have to gain IP = 6.6 eV in going from the gas to the liquid phase in order to account for the experimental photoionization. The gain would be due in part to the dielectric effect which makes it easier to remove an electron from a given positive charge and in part to a _partial_ solvation of the positive ion. If we estimate a

3 eV gain by the change in dielectric constant of the medium, we need only a <u>partial</u> solvation of H_2O^+ of about 3.5 eV to facilitate the ionization of H_2O^*. It should be emphasized that only partial solvation energy becomes available by this mechanism because complete solvation, which is exothermic by about 5 eV, would require the rearrangement of water molecules around H_2O^*; this could hardly take place within 10^{-13} sec, the estimated lifetime of H_2O^*. Our conclusion that H_2O^*, produced as intermediate, dissociates or ionizes within 10^{-13} sec without any substantial atomic rearrangement of the solvent matrix could be supported by the fact that the ratio ionization/dissociation does not change with temperature.[6] If rearrangement of water molecules around the nascent H_2O^+ is a prerequisite for its formation, a significant temperature effect would be expected between 30° and 73°C. It is difficult to calculate the polarization energy involved in the hydration of H_2O^+, but it is expected to be a major fraction of the hydration energy just as has been shown for the case of hydration of electrons in water.[20]

The effect of the dielectric constant on the ionization potential in MeOH and the polarization energy around the nascent $MeOH^+$ in methanol are unknown. It is unlikely, however, that their sum is sufficient to allow the photoionization of liquid MeOH with merely 4.7 eV. Thus the photoionization of methanol calls for a different explanation of the photoionization both in water and in methanol.

It is suggested that the driving force of the photoionization are the proton affinities of H_2O and MeOH which amount to 7.15 [30] and 7.8 [32] eV respectively. If an excited water molecule H_2O^* with an excitation energy of 6.05 eV transferred a proton to an adjacent water molecule <u>without</u> any rearrangement of the solvation shell of either,

$$H_2O^* + H_2O \longrightarrow (OH^-)^* + H_3O^+$$

the process would result in an excess of 7.15 + 6.05 − 5.16 > 8.0 eV. The remaining OH^- ion is thus left with an excitation energy which is much higher than its electron affinity (EA(OH) = 1.89 eV)[33] and which results in its immediate ionization.

$$(OH^-)^* \longrightarrow OH + e^-$$

The released electron will undergo hydration to form e^-_{aq} only at a later stage. The proton transfer reaction is a very fast process[30] $t_{\frac{1}{2}} \sim 10^{-13}$, comparable in rate to the dissociation step

$$H_2O^* \longrightarrow H + OH.$$

The analogous reactions in methanol will follow a similar pattern where the reaction $MeOH + MeOH^* \longrightarrow MeO^- + MeOH_2^+$ is exothermic by $4.7 + 7.8$ [32] $- 4.45$ [34] > 8 eV. Like OH^- also, MeO^- is left excited and, in view of its low electron affinity (EA(MeO) = 0.39 eV),[35] it undergoes immediate ionization:

$$(MeO^-)^* \longrightarrow MeO + e^-$$

In spite of the favorable energetics, methanol undergoes less extensive photoionization than water. The ratio e^-/H in MeOH is 1/30 compared to 1/7 in water. This difference may be explained by the less favorable orientation in the liquid phase which does not allow every hydroxylic H to be in the optimal distance from an adjacent oxygen. Furthermore, the weaker C-H and O-H bond strengths in methanol (4.05 and 4.45 eV, respectively)[34] facilitate dissociation rather than ionization. The same trend is also observed in the radiolysis of methanol[36] where the e^-/H is 1:2.2 compared with 5.4:1 in water.[20]

The proposed mechanism differs from that of Boyle et al.[6] by not requiring any solvent molecule rearrangement which would take $>10^{-11}$ sec; thus solvated H_3O^+ or e^- are not produced as primary products in our mechanism. In principle one could expect our mechanism to occur also in the gas phase if H_2O^* would survive until it collides with a second water molecule. As the lifetime of H_2O^* is $\sim 10^{-13}$ sec, photoionization in the gas phase by low energy photons would occur only at pressures above 100 atm.

In conclusion, it may be stated that any associated protonic liquid might undergo photoionization at much lower energies than the ionization potential in the gas phase. This is facilitated by the hydrogen-bonded structure in conjunction with proton affinity. The preorientation of the excited proton donor and the proton acceptor are critical so that the proton may be transferred within the duration of a single vibration of the excited molecule. Only under these conditions can the proton transfer compete with homolytic dissociation.

REFERENCES

1.   J. W. Weeks, G. M. A. C. Meaburn, and S. Gordon, Radiation Res. 19, 559 (1963).

2.   P. D. Stevenson, J. Phys. Chem. 69, 2145 (1965).

3.   L. R. Painter, R. D. Kirkhoff, and E. T. Akarawa, J. Chem. Phys. 51, 243 (1969).

4.   J. Barrett and J. H. Baxendale, Trans. Faraday Soc. 56, 37 (1960).

5.   U. Sokolov and G. Stein, J. Chem. Phys. 44, 3329 (1966).

6.   J. W. Boyle, J. A. Ghormley, C. J. Hochanadel, and J. F. Riley, J. Phys. Chem. 73, 2886 (1969).

7.   M. S. Matheson, W. A. Mullac, and J. Rabani, J. Phys. Chem. 67, 2613 (1963).

8.   N. Getoff and G. O. Schenck, Photochem. and Photobiol 8, 167 (1968).

9.   E. L. Chaney, L. G. Christophorou, P. M. Collins, and J. G. Carter, J. Chem. Phys. 52, 4413 (1970).

10. G. R. A. Johnson and M. Simic, J. Phys. Chem. 71, 2775 (1967).

11. J. M. Warman, K. D. Asmus, and R. H. Schuler in Radiation Chemistry II, Am. Chem. Soc. Adv. Chem. 82, 25 (1968).

12. K. D. Asmus and J. H. Fendler, J. Phys. Chem. 72, 4285 (1968).

13. T. Lui, G. Moe, and A. B. F. Duncan, J. Chem. Phys. 19, 71 (1953).

14. E. D. Nostrand and A. B. F. Duncan, J. Am. Chem. Soc. 76, 3377 (1954).

15. L. H. Friedman, J. Am. Chem. Soc. 76, 3294 (1954).

16. R. M. Noyes, Prog. in Reaction Kinetics, G. Parker, Ed., Vol. 1, p. 129, Pergamon Press, 1961.

17. M. Anbar, D. Meyerstein, and P. Neta, J. Chem. Soc. [B], 742 (1966).

18. M. Anbar and P. Neta, Int. J. Appl. Rad. Isotopes 18, 493 (1967).

19. M. S. Matheson and J. Rabani, J. Phys. Chem. 69, 1324 (1965).

20. E. J. Hart and M. Anbar, "The Hydrated Electron," Wiley 1970, Ch. V.

21. J. K. Thomas, J. Rabani, M. S. Matheson, and E. J. Hart, J. Phys. Chem. 70, 2409 (1966).

22. B. de B. Darwent, Bond Dissociation Energies in Simple Molecules, NSRDS-NBS #31, 1970.

23. J. L. Franklin et al., Ionization Potentials of Gaseous Positive Ions, NSRDS-NBS #26, 1969.

24. D. P. Stevenson in "Structural Chemistry and Molecular Biology," edited by A. Rich and N. Davidson, W. H. Freeman, Publisher, San Francisco, 1968.

25. B. Brehm, Z. Naturfursch. 21a, 196 (1966).

26. W. Kaye and R. Poulson, Nature 193, 675 (1962).

27.   M. I. Al Joboury and D. W. Turner, J. Chem. Soc. 4434 (1964).

28.   C. Vermeil, M. Matheson, S. Leech, and F. Muller, J. Chim. Phys. 61, 596 (1964).

29.   C. Vermeil, F. Muller, M. Matheson, and S. Leech, Bull. Soc. Chim. Belg. 71, 837 (1962).

30.   J. B. E. Randles, Trans. Faraday Soc. 52, 1573 (1956).

31.   M. S. B. Munson, J. Am. Chem. Soc. 87, 2332 (1965).

32.   R. A. Strechlow, private communication.

33.   J. Kay and F. M. Page, Trans. Faraday Soc. 62, 3081 (1966).

34.   S. Benson, J. Chem. Educ. 42, 502 (1965).

35.   G. Freeman, Rad. Res. Revs. 1, 1 (1968).

36.   G. E. Adams and R. O. Sidgwick, Trans. Faraday Soc. 60, 865 (1964).

SOLUTE INTERFERENCE EFFECTS IN FREEZING POTENTIALS OF DILUTE ELECTROLYTES

Gerardo Wolfgang Gross

New Mexico Institute of Mining and Technology

Socorro, New Mexico 87801

Abstract

Charge separation at an advancing ice/water interface is caused by preferential incorporation of ion constituents of one sign into the solid phase. Magnitude and sign of the potential difference depend on concentrations and kinds of solute species. Electrical neutrality is restored by a flux of hydroniums (protons) and hydroxyls, present in solution or generated by electrolysis. Interference effects occur between competing solute species in a freezing solution. The interference is positive if it enhances, negative if it reduces the freezing-potential difference. A reversal of the potential sign was observed in certain cases. A large excess of hydrogen or hydroxyl ions in the original solution always interferes negatively. Ammonium gave rise to large freezing potentials (ice positive) in concentrations as low as 10^{-10}M, provided dissolved CO_2 was in some excess. Similar effects by CO_2 were observed with ammonium carbonate and bicarbonate solutions. NaCl (10^{-5} M to 10^{-6} M) negatively interfered with ammonium carbonate or bicarbonate concentrations lower than 10^{-6}M. NaCl by itself gave high freezing potentials (ice negative) but dissolved CO_2 (absorbed from air) caused strong negative interference. Ionic interference effects are important for an understanding of ice electrification. They may increase or decrease both the concentration range in which charge separation occurs with a given solute and the amplitude of the potential differences. Thus, they explain some of the scatter observed in this type of laboratory measurement. They could be useful in assessing the effect of the interfacial electric field on solute distribution coefficients but the question whether or how these electrochemical

interference effects relate to the structure of the phase boundary
(or of the bulk phases in the vicinity of the boundary) cannot at
present be answered.

PURPOSE

Ice or "ice-like" water play a role in many natural processes,
physical, chemical, and biological. Specifically, the appearance
of the ice phase is correlated with the onset of certain processes
of cloud electrification. Ice may be conceived as participating
in these phenomena in a number of ways. Electrical charge may be
separated simultaneously with and as an integral part of the freezing process (graupel formation, riming) in clouds. Furthermore,
ice in clouds may store and/or conduct charge induced by other
processes (atmospheric ionization, corona discharge).

Since Faraday (1845) processes of charge separation in ice
have been studied by many investigators and a number of mechanisms
have been proposed (Israel, 1964), but it has not been possible to
pinpoint whether one or a combination of them actually is responsible for cloud processes. Atmospheric phenomena, because of
their complexity, are very difficult to reproduce in the laboratory.
But even experimental data are poorly reproducible from one inves-
tigator to another.

The present research had for an objective the experimental
and quantitative study of only one electrification process, viz.,
the non-equilibrium electrochemical charge separation at the advancing boundary of an ice particle growing in water (Workman-
Reynolds effect), attributed to trace solutes (Workman and Reynolds,
1950).

Initially, the research effort was directed toward a system-
atic study of a large number of (mostly inorganic) pure laboratory
reagents as a function of solute species, concentration, freezing
rate, and pH (Gross, 1965; Cobb and Gross, 1969). The largest
charge separations were observed in the alkali fluorides, the
highest freezing potentials in ammonium carbonates and bicarbon-
ates. However, in ice frozen from these last two compounds a large
blocking potential was found to develop at the ice electrode
(LeFebre, 1970). This blocking potential prevents true interface
charge separation measurements. Because of the ubiquitous presence
of ammonium and bicarbonate in natural and especially in atmospheric waters (Junge, 1963; Garrels and MacKenzie, 1967; Gibbs, 1970),
this point is worth emphasizing.

One single reagent grade salt, acid, or base was used in most
of the laboratory tests discussed in the literature up to now.

Fig. 1: Carbon-dioxide free chamber in which the interference
experiments are done. A detailed description is found in
Gross (1971). The freezing cup is shown on the sheathed
cold block. It is being filled with carbon-dioxide free
high-purity water. Vertically above it is the stainless-
steel shield which contains the water electrode, a spiral
of platinum wire soldered into a coaxial connector. Bat-
tery-operated pH meter is in left foreground. Sealed
admission port visible in right background.

Natural waters usually contain a suite of solutes in varied proportions. They react with the atmosphere mainly by carbon dioxide
exchange (and by dissolving air). This report presents results of
testing mixtures of solute species that are either of particular
laboratory interest or are associated in natural waters.

EXPERIMENTAL PROCEDURES

All work was carried out in a glove box in which carbon-dioxide free air was circulated (Fig. 1). The procedures and equipment have been described in detail elsewhere (Gross, 1971). Potential differences are those of ice with respect to water.

MECHANISM OF CHARGE SEPARATION

The growing ice particle rejects the bulk of impurities in the
water. The advancing phase boundary thus functions as a semi-
permeable membrane in reverse osmosis. Superposed on this whole-
sale rejection is the electrochemical charge separation, which is
ascribed to small differences in the incorporation (into the ice
phase) of the solute anion and the solute cation constituents.
For the present purpose, an ion constituent is an element or rad-
ical entering into the composition of an electrolyte and forming a
cation or an anion when the electrolyte is dissolved in water at
room temperature. Typically in the alkali and ammonium fluorides
the anion constituent is incorporated preferentially. In other
ammonium salts (carbonate, bicarbonate, chloride), some lead salts
(acetate, nitrate), and sulfates and acetates of sodium and potas-
sium, however, the cation constituent is preferred. The analytical
difference between solute cation and anion constituents in the ice
phase decreases with increasing freezing rate. At the largest
freezing rates, at which also the highest freezing potentials were
observed, the difference may be of the order of a few percent, un-
detectable by the analytical methods used.

During the differential incorporation process, charge and
chemical balance are restored (or maintained) by hydrogen and
hydroxyl ions, respectively. This explains the great sensitivity
of the process to solution pH.

In all solutes studied up to the present, the charge on the
ice has the same sign as the preferred solute ion constituent. On
the basis of the sign of ice charge and size of charge separation,
solutes were classified into three types (Table I).

For chemical and electrical balance to be maintained at any
given moment, the instantaneous flux densities of solute constituents across the advancing phase boundary into the ice must obey

TABLE I. CLASSIFICATION OF SOLUTES (MODIFIED AFTER COBB AND GROSS, 1969).

Type	Chemistry	Ice Charge Sign	Neutral. by	Potential*	Charge Sep.**	pH***	Freez.Rate Dependence	Conc. Range
I	Alkali halides, NH_4F	(-)	H^+	Moderate	Large	7+	Large	10^{-6} to $10^{-4}\underline{M}$
II	Salts of NH_4^+, Pb^{++}; sulfates; nitrates; oxalates; acetates.	(+)	OH^-	Large	Small****	7-	Small	10^{-10} to $10^{-3}\underline{M}$
IIIa	Acids, atm. CO_2	0	None	0	0	-	-	All
IIIb	Bases (carbonate-free, solub. hydroxide).	0	None	0	0	-	-	All

*   Measured with 10^{14} ohms input impedance.

**   Measured with 10K ohms input impedance.

*** Initial solution, for largest potential difference.

**** Electrode blocking potential prevents measurement.

the relationship

$$z_r^+ v_r^+ k_r^+ C_r^+ + v_H C_H = z_p^- v_p^- k_p^- C_p^- \qquad (1a)$$

for Type I solutes, and

$$z_r^- v_r^- k_r^- C_r^- + v_{OH} C_{OH} = z_p^+ v_p^+ k_p^+ C_p^+ \qquad (1b)$$

for Type II solutes, where:

V = velocity at which ion constituents cross the phase boundary (function of solute species and perhaps concentration, freezing rate, phase boundary structure, and electric phase boundary field).

Superscripts (-) and (+) refer to solute anions and cations, respectively.

Subscript p designates the preferentially incorporated solute ion constituent, subscript r the preferentially rejected species.

C = equivalent solute concentration at the phase boundary (function of initial solute concentration, solute concentration gradient at the phase boundary, diffusion coefficient, and partition coefficient).

C_H, C_{OH} are the equivalent concentrations of the neutralizing species, that is, hydroniums (protons) in Type I and hydroxyls in Type II solutions.

Z = ion valency.

k = partition coefficient (function of freezing rate, concentration, solute species, and electric interface field).

Equations (1) assume that the effective charges equal the nominal charges of each carrier species.

Charge separation is possible only if the fluxes on the left side of the equation lag with respect to those on the right side. This is likely to occur at high freezing rates or when the neutralizing species (H^+ or OH^-, respectively) is scarce. Some limiting cases of particular interest are shown in Table II; they serve as a qualitative frame of reference for the discussion of interference effects. Note, in particular, that a high interface field is not conducive to complete differential ion separation. A phenomenological analysis has been made by B. Gross (1954) and by LeFebre (1967).

TABLE II. QUALITATIVE DEPENDENCE OF ICE SOLUTE CONTENT, FREEZING POTENTIAL, AND FRACTIONAL ION CONTENT (MEASURE OF ION SEPARATION) ON CONCENTRATIONS OF INITIAL SOLUTION AND FREEZING RATE. FOR EXPLANATION OF SYMBOLS, SEE EQ. (1).

| Freez. Rate | EQUIVALENT CONCENTRATION | | | | | Freez. Potential | Fractional Ion Content Ice $C_p/(C_r + C_p)$ |
| | Initial Solution | | Growing Ice | | | | |
	Solute	C_H or C_{OH}	C_r	C_H or C_{OH}	C_p		
Small*	low**	low	$\rightarrow 0$	$= C_p$	low	low	$\rightarrow 1$
Small	high***	low	$< C_p$	low to v.l.	moderate	0	0.5 – 1
High****	low to mod.	low	$< C_p$	low	consid.	high	0.5 – 0.6
High	high	low	$= C_p$	neglig.	large	$\rightarrow 0$	0.5
Any	any	high	unknown	consid.	unknown	0	unknown

*   typically < 1 x 10^{-4} cm/sec.

**   typically < 10^{-5} $\underline{M}$

**   typically > 10^{-3} $\underline{M}$

****   typically > 5 x 10^{-4} cm/sec.

Charge separation has been discussed as though this property for any given species were independent from other ion constituent species present in solution. Actually, however, they are coupled. The same ion constituent may be more or less acceptable (to the ice phase) depending on the other species. An extreme case is ammonium, which is preferentially incorporated when it is coupled with bicarbonate, but which is preferentially rejected in ammonium fluoride. Using partition coefficients, it appears that, qualitatively

$$k_1^- > k_1^+ \qquad \text{in } NH_4F \text{ solutions}$$

$$\tag{2}$$

$$k_2^+ > k_2^- \qquad \text{in } NH_4CO_3 \text{ solutions,}$$

where subscripts 1 and 2 refer to one of the two solutes as shown in the equations. Moreover,

$$\frac{k_1^+}{k_2^+} \gg 1. \tag{3}$$

Other cases of coupling have been described in the literature (e.g. : Pruppacher, Steinberger, and Wang, 1968; Uzu and Sano, 1970). Effects were correlated with electronegativity and ionic radius.

INTERFERENCE MECHANISMS

Solute interference effects during electrochemical charge separation by freezing were investigated experimentally and some were predicted on the basis of the considerations discussed in the preceding section. A classification of these effects together with selected experimental data are presented in this section. The list mainly reflects this writer's experience and does not claim to be exhaustive.

(1) Interferences affecting the concentration of the "neutralizing species" (H^+, OH^-, see Eq. 1). This is interference of a Type III with a Type I or a Type II solute.

(a) The neutralizing species is suppressed. Freezing potential peak (charge separation) is increased. Interference is positive. Cases recognized: atmospheric carbon dioxide (Fig. 2) or dilute acids (Fig. 3) in Type II solutions (pH depressed below 7). Soluble hydroxide bases (NaOH, KOH, NH_4OH) in Type I solutions (pH increased above 7), Fig. 4. Fig. 5 compares freezing potentials measured in reagent-grade solutions of ammonium carbonate and bicarbonate and, though not a case of interference, illustrates the

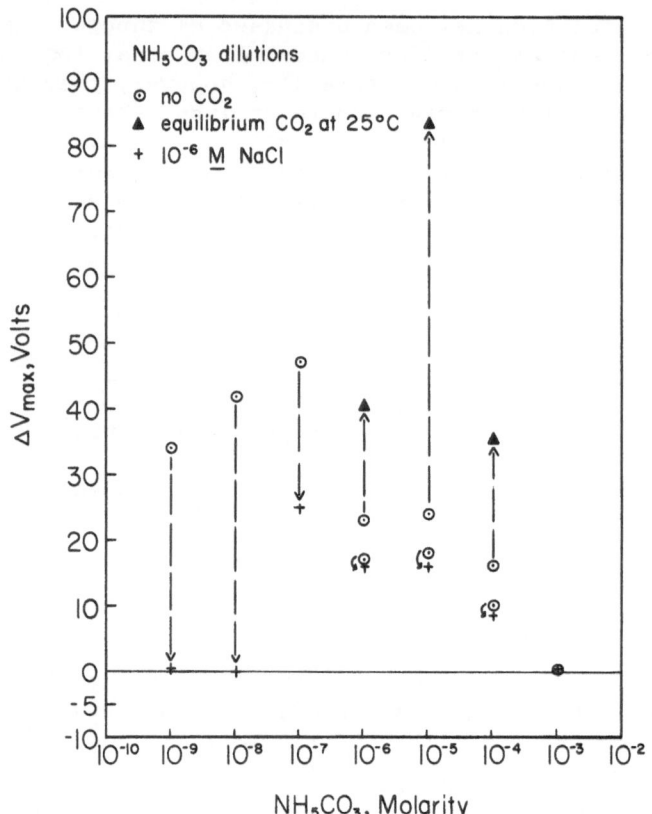

<u>Fig. 2</u>: Positive and negative interference effects in ammonium
        bicarbonate solutions (arrows indicate direction of
        interference).

principle that the lower relative concentration of the neutralizing
species (in this case hydroxyl, controlled by dissociation of the
pure reagent itself) gives the higher charge separation.

(b) The neutralizing species is <u>augmented</u>. Freezing potential (charge separation) is decreased. Interference is <u>negative</u>.
<u>Cases recognized</u>: Atmospheric carbon dioxide (Figs. 4, 6) or
dilute acids in Type I solutions (pH depressed below 7); soluble
hydroxide bases in Type II solutions (pH increased above 7), Fig.
3.

(2) Interference of a Type IIIa with a Type IIIb solute.
This actually creates a Type I or a Type II solute, except that
proportions of the solute anion and cation constituents may be

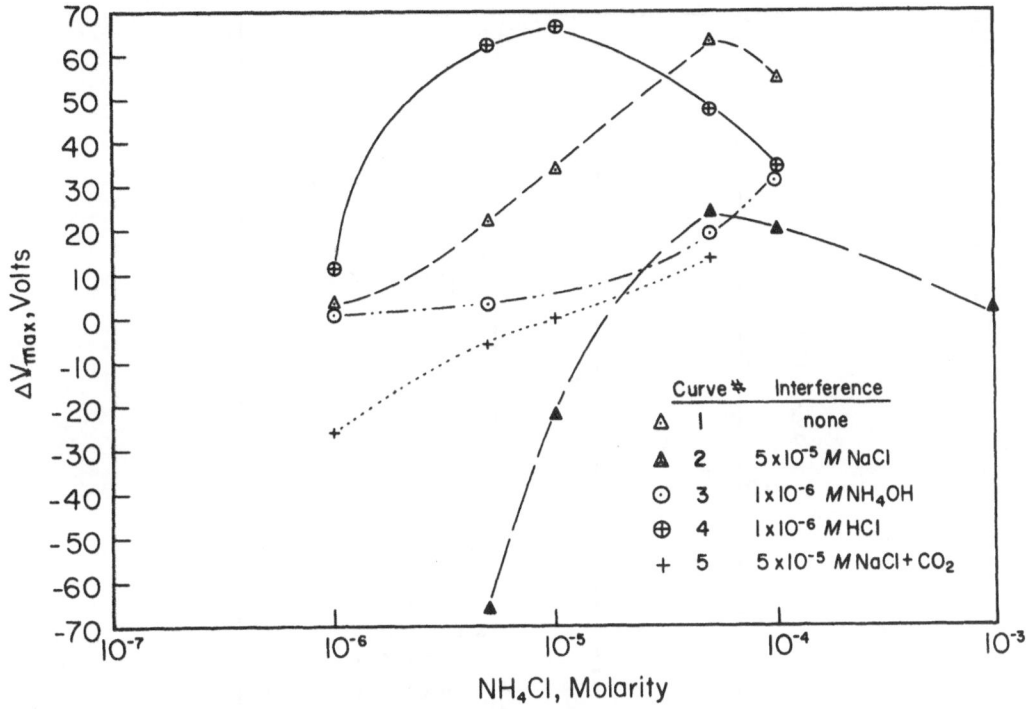

Fig. 3: Positive and negative interference effects in ammonium chloride solutions. When carbon dioxide is present in addition to NaCℓ, the latter's effect is partially inhibited (compare curves 2 and 5).

non-stoichiometric.

(a) Mixtures of reagent-grade acids with bases.

(b) Interference of atmospheric carbon dioxide with a reagent-grade, soluble hydroxide base (NaOH, NH_4OH). Figs. 7, 8. Very small concentrations of ammonia, in the presence of an excess of carbon dioxide, may achieve sizable freezing potential effects (Fig. 8).

(c) Water distilled into glass (Pyrex) or polypropylene and then highly purified by ion exchange, frequently exhibited high, poorly reproducible freezing potentials of Type II. This effect was greatly enhanced by storage (especially in glass but also sometimes in polypropylene or Teflon), and by exposure to air or dry ice in a vacuum. Appreciable amounts of silica (0.1 ppm) were found in some such samples of water exhibiting high potentials. Dissolution of silicic acid is promoted by carbon dioxide (Garrels and MacKenzie, 1967).

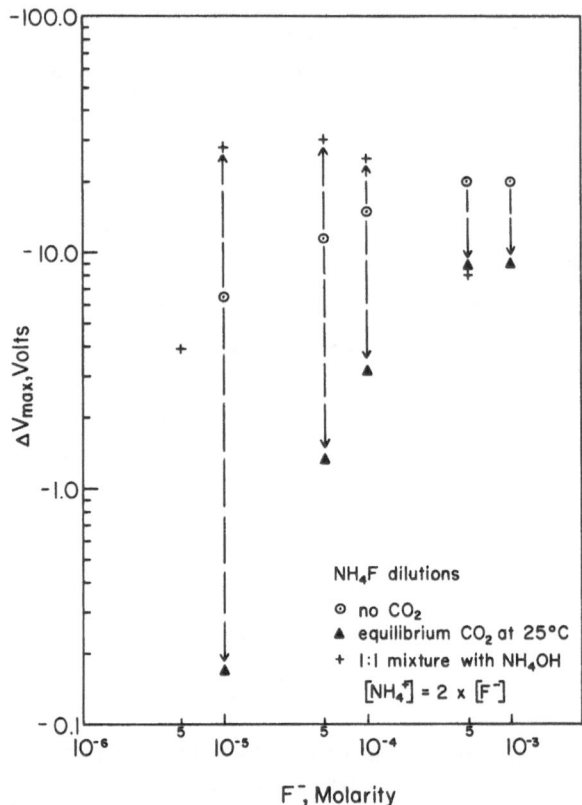

ΔV_{max}, Volts

F⁻, Molarity

NH₄F dilutions

⊙ no CO₂
▲ equilibrium CO₂ at 25°C
+ 1:1 mixture with NH₄OH

$[NH_4^+] = 2 \times [F^-]$

Fig. 4: Positive and negative interference effects in ammonium
fluoride solutions. Slightly negative interference of
5×10^{-4} $\underline{M}$ NH₄F with equal concentration of NH₄OH is
probably a concentration effect.

(3) Interference between Type I and Type II solutes. The
coupling of partition coefficients of solute anion and cation constituents is changed by the interfering solute.
        (a) Interference causes a <u>concentration</u>-dependent rever-
sal in the sign of the charge separation (Fig. 3, interference of
NaCℓ with NH₄Cℓ).
        (b) Interference causes a concentration-dependent and/
or a <u>time</u>-dependent reversal in the sign of the charge separation.
(Figs. 9 and 10). For certain proportions and concentrations of
the interfering solutes the freezing-potential curve exhibits two
maxima of comparable size and of opposite sign as the freezing
progresses (Fig. 10). Thus far, this type of interference has only
been observed in the system NH_4^+, F^-, HCO_3^-, or its equivalent
NH₄OH, HF, CO₂ (atmospheric).

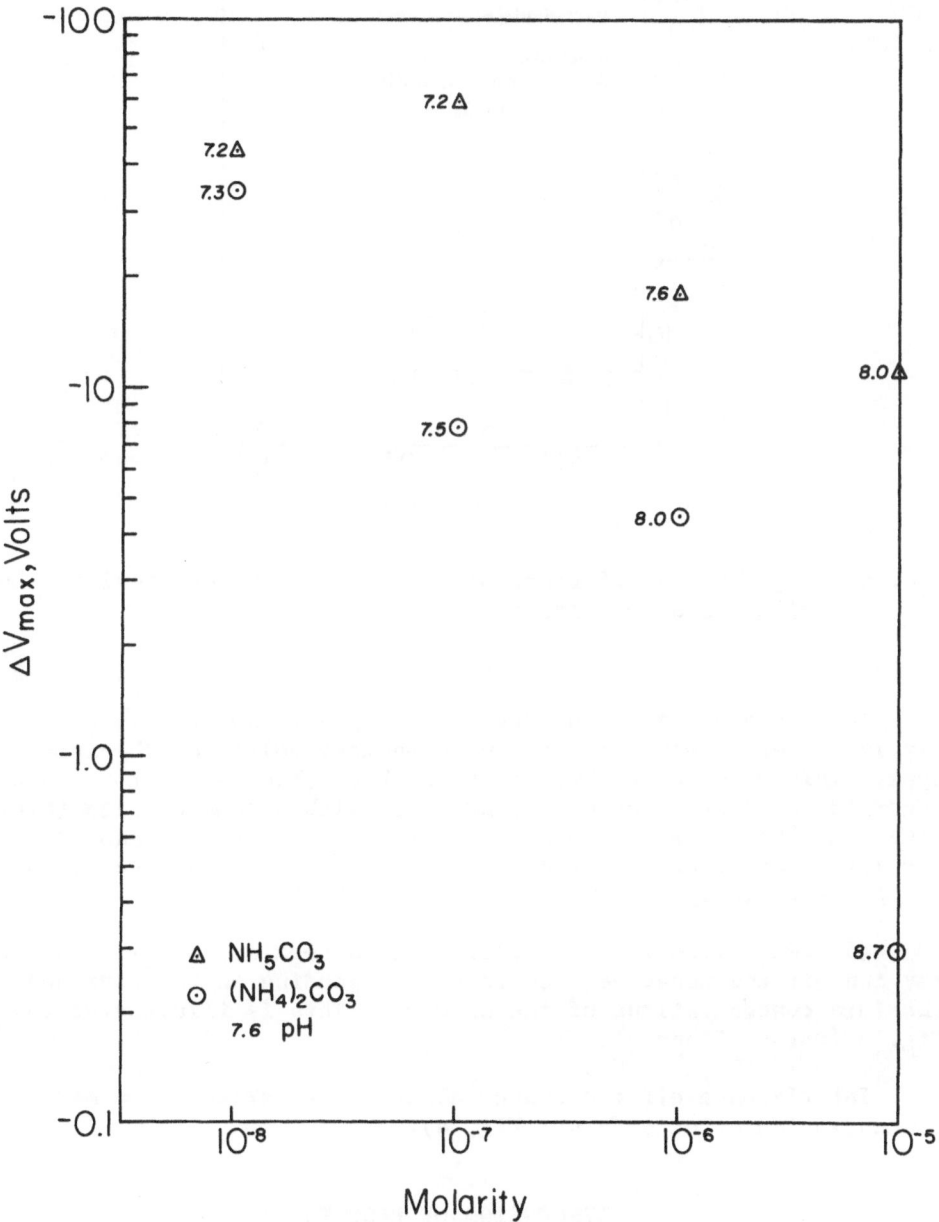

<u>Fig. 5</u>:   Maximum freezing potentials of dilute ammonium carbonate
and bicarbonate solutions. For a given concentration,
the bicarbonate exhibits the higher potential because of
lower pH.

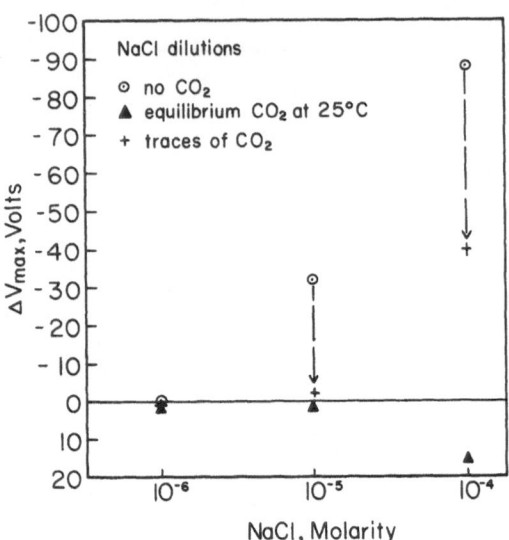

Fig. 6: Negative interference of atmospheric carbon dioxide with
          dilute sodium chloride solutions.

(4) Theoretically, a Type I or a Type II solute may positively or negatively interfere with another solute of the same
type. This case has not yet actually been observed. It has been
reported, however, that different salts with a common anion (e.g.
CsCl, KCl, NaCl) or a common cation (e.g. KI, KCl, KF) show different levels of freezing-potential differences. Hence interference
effects are likely (Pruppacher et al, 1968; Uzu and Sano, 1970).

(5) When Type I and Type II solutes coexist in solution, CO_2
may inhibit the negative interference depending on relative and
absolute concentrations of the solutes. This is illustrated in
Fig. 3 (curves 2 and 5).

Interference effects cannot always be separated from mere
concentration effects (e.g.: Fig. 4).

DISCUSSION OF RESULTS

The investigated interference effects further confirm the
basic feature of the postulated mechanism of charge separation,
namely, the neutralization of preferentially incorporated ion
constituents by H^+ or OH^-, respectively (Workman and Reynolds,
1950; Cobb and Gross, 1969).

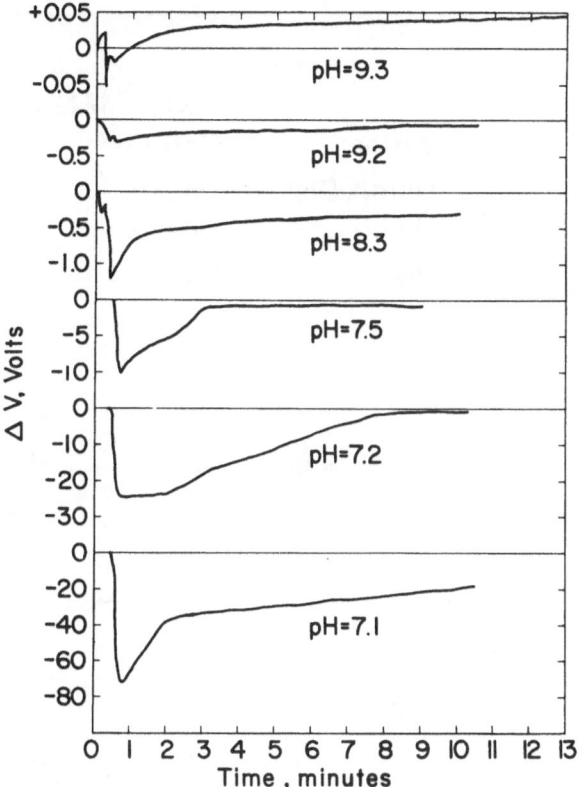

<u>Fig. 7:</u>   Curves of freezing potential vs. time for an ammonia
         solution (5 x 10⁻⁵ M) repeatedly exposed to air. Redrawn
         from the actual records. A negative potential deflection
         means a positive ice charge because the ice electrode was
         grounded. Note that the potential scale is different for
         each curve.

    They also supply a plausible mechanism for both concentration-
dependent and time-dependent voltage reversals (Reynolds, Brook
and Gourley, 1957; Parreira and Eydt, 1965; Gross, 1968, 1970;
Workman, 1969; Murphy, 1970). Reynolds et al. (1957) induced time-
dependent voltage reversals by replacing solutes during freezing.
The present work (also Gross, 1971) shows that voltage reversals
can occur, too, if fluoride, ammonium, and bicarbonate ions coexist
in the initial mother solution such that fluoride is in 3 : 1 to 6
: 1 excess over bicarbonate. In every case, the first potential
peak was negative (Fig. 10) due to the preferential incorporation
of fluoride (Type I). As a consequence, its concentration in the
interface region becomes depleted relative to ammonium and

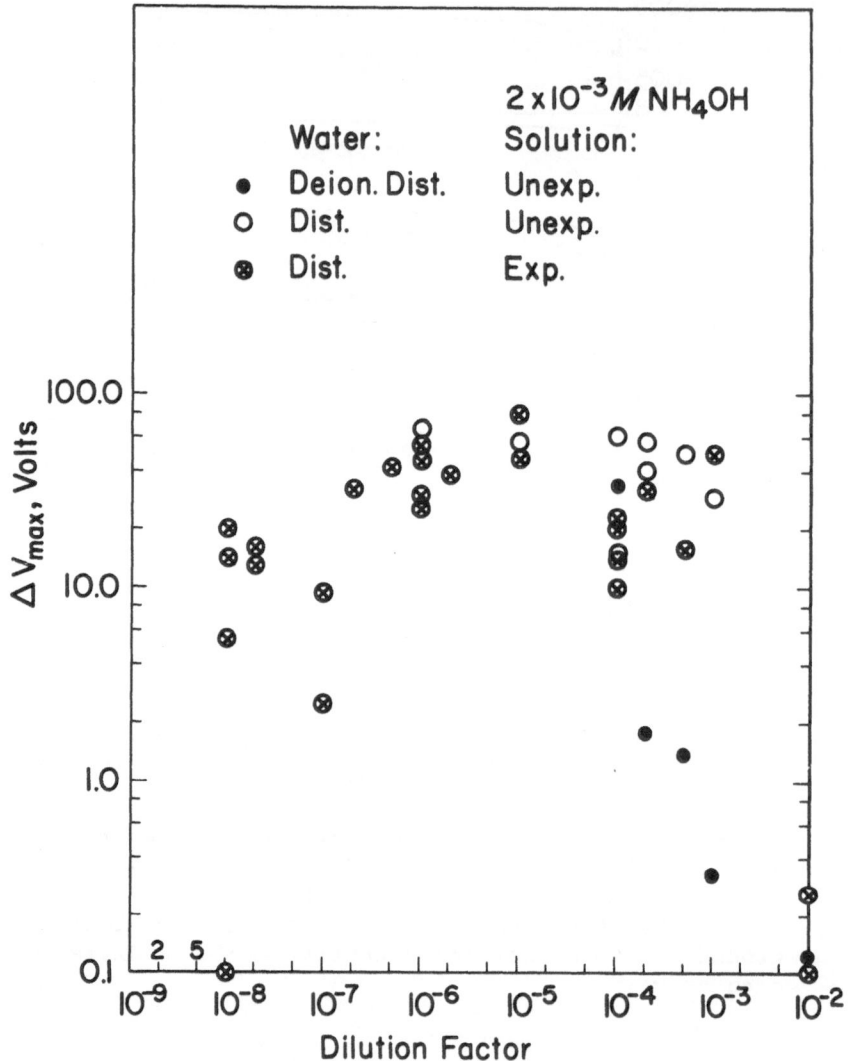

<u>Fig. 8</u>: Positive interference of carbon dioxide with ammonia.
Maximum freezing potential differences in ammonia solutions as a function of dilution and CO_2 content. (Ammonia
concentration in moles per liter is 2×10^{-3} multiplied
by dilution factor). Deionized distilled water is free
of CO_2. Distilled water contains traces of CO_2, actually
in excess of ammonia at dilution factors smaller than
about 10^{-4}, so that it does not matter whether the stock
solution ($2 \times 10^{-3}\underline{M}$) had been exposed or not to air.
Notice the unusually broad concentration range in which
this positive interference effect was observed.

bicarbonate so that the predominant solute incorporation (and the potential sign) must change (see Eqs. 2 and 3). Once established, the Type II mode is stable in this system.

Parreira and Eydt's potential reversal in NaCl solutions from Type I to Type II as the concentration dropped below 5×10^{-6} M (Parreira and Eydt, 1965, Fig. 3) may be a result of interference. Murphy (1970) has taken up a suggestion by Parreira and Eydt and explained such reversals as caused by an intrinsic polarization process forming "trees" of oriented dipoles at or near the phase boundary in the freezing of pure water. These dipole chains relax as the interface advances, thus giving rise to a charge transfer. He assumes explicitly that water with very small impurity traces (less than 10^{-7} M) can be considered to be "pure". Our results with ammonia in the presence of carbon dioxide show that certain solutes may produce specific electrical effects at such dilutions (Fig. 8). The theoretical importance that attaches to possible mechanisms of charge separation in the freezing of pure water, as suggested by Parreira and Eydt and by Murphy requires further investigation of this problem by techniques considerably more refined than those used up to now. This question acquires additional significance in view of the experimental observation of static multilayers of oriented water dipoles at the walls of living cells (Cope, 1971). Can a dynamic system, e.g. the interface of an ice particle growing in water, give rise to similar dipole layers that are continuously renewed as the interface progresses?

What is the effect of the interfacial electrical field on the partitioning of ionized solutes? It is believed to be small (LeFebre, 1967), for the preferentially incorporated ion constituent in any case (Gross, 1965). The question is, however, of fundamental importance for an understanding of the structure of the ice/water interface and should be investigated thoroughly. Avenues of approach might lie in the use of an internal shunt (Gross, 1965) or in the use of solute interference such that the electric interface field is completely suppressed.

Earlier work seemed to indicate that freezing potentials require very pure reagents and narrow concentration ranges for their development. This would severely restrict their applicability to natural waters (Workman, 1969). The present results suggest that certain interference processes can relax these requirements by broadening the concentration range in which appreciable charge separation may occur, by changing the sign of the potential peak, or by increasing its amplitude. Ammonia, carbon dioxide, and possibly silica figure prominently in the interference processes discussed in this paper. These solutes also are abundant and universal in natural waters (Junge, 1963; Garrels and MacKenzie, 1967; Gibbs, 1970). Much work remains to be done in clarifying the nature

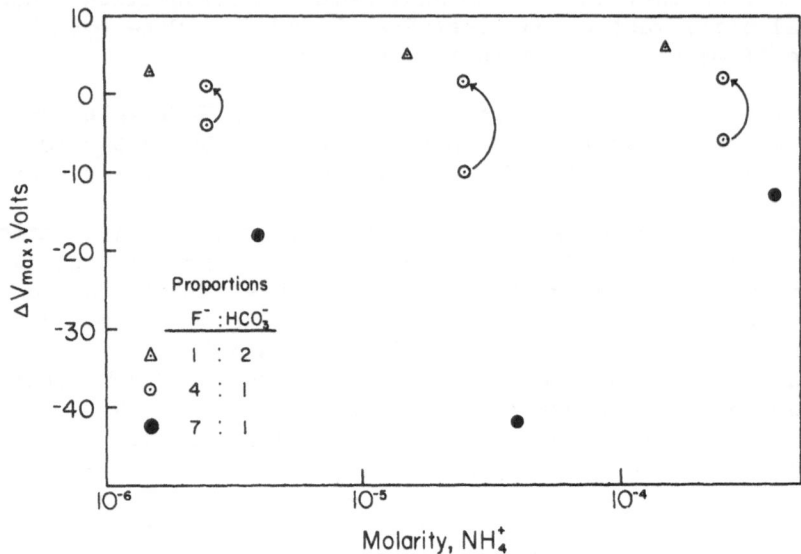

Fig. 9: Interference of ammonium fluoride and ammonium bicarbonate. Time-dependent reversal (arrows) of the potential
peak in a narrow range of proportions (F/HCO$_3$ = 3/1 to
6/1). Potential differences are small in the range in
which this reversal was observed. See also Fig. 10.

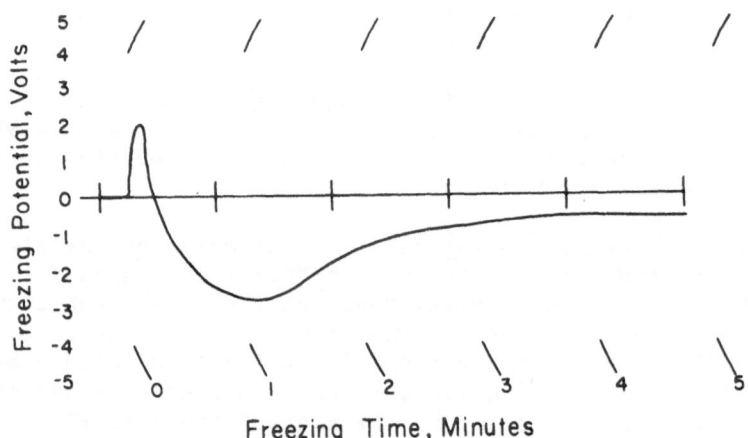

Fig. 10: Freezing potential vs. time for a mixture of 4 x 10^{-4} $\underline{M}$
NH$_4$F and 1 x 10^{-4} $\underline{M}$ NH$_5$ CO$_3$. Redrawn from record. Potential sign convention is the same as in Fig. 7.

of interference processes as well as their role in natural electrification.

SUMMARY AND CONCLUSIONS

Electrochemical charge separation in freezing water (Workman-
Reynolds effect) is modified in the presence of multiple solute
species. This interference may either increase or decrease the
charge separation. One important type of interference alters the
solution pH. Depending on the type of solute, a moderate increase
of pH above 7 or a moderate decrease below 7 may result in maximum
charge separation effects, and conversely, in their suppression.
Another type of interference causes a reversal of the freezing-
potential sign as a function of concentration (of the interfering
species) or of concentration <u>and</u> time. This effect probably
operates through the solute partition coefficients. The most
important interference effects studied thus far involve ammonia,
carbon dioxide, and possibly silica. Interference effects cannot
always be separated from concentration effects. Certain interference processes broaden the concentration range in which appreciable charge separation may occur for a given solute.

ACKNOWLEDGMENTS

Mrs. Nancy Smith Frame performed the experimental work and
Miss Lynn Isham the silica analyses. This work was supported by
the Office of Naval Research under contracts Nonr-815(05) NR 082-
094, and N00014-67-A-0267-0005 NR 082-094/9-29-70(412).

REFERENCES

Cobb, A. W., and G. W. Gross (1969), Interfacial electrical effects
observed during the freezing of dilute electrolytes in water.
J. Electrochem. Soc. <u>116</u>, 796-804.

Cope, F. W. (1971), Structured water and complexed Na^+ and K^+ in
biological systems. This Symposium.

Garrels, R. M., and F. T. MacKenzie (1967), Origin of the chemical
compositions of some springs and lakes. Adv. in Chem. Series
<u>67</u>, 222-242.

Gibbs, R. J. (1970): Mechanisms controlling world water chemistry.
Science <u>170</u>, 1088-1090.

Gross, B. (1954): Theory of Thermodielectric effect. Phys. Rev.
<u>94</u>, 1545-1551.

Gross, G. W. (1965): The Workman-Reynolds effect and ionic transfer processes at the ice solution interface. J. Geophys. Res.
70, 2291-2300.

Gross. G. W. (1968): Some effects of trace inorganics on the ice/
water system. Adv. in Chem. Series 73, 27-97.

Gross, G. W. (1971): Freezing potentials in the system $H_2O - NH_3 -
CO_2$ at controlled concentrations. J. Atm. Sci., Sept. '71.

Israel, H. (1964): Probleme der Gewitterforschung.
Forschungsberichte des Landes Nordrhein-Westfalen, 1408.

Junge, Ch. E. (1963): Air Chemistry and Radioactivity. Academic
Press, New York.

LeFebre, V. (1967): The freezing potential effect. J. Coll. and
Interface Sci. 25, 263-269.

LeFebre, V. (1970): The pseudo melting potential effect. J. Coll.
and Interface Sci. 33, 572-577.

Murphy, E. J. (1970): The generation of electromotive forces
during the freezing of water. J. Coll. and Interface Sci. 32,
1-11.

Parreira, H. C., and J. A. Eydt (1965): Electric potentials
generated by freezing dilute aqueous solutions. Nature 208,
33-35.

Pruppacher, H. R., E. H. Steinberger, and T. L. Wang (1968): On
the electrical effects that accompany the spontaneous growth of
ice in supercooled aqueous solutions. J. Geophys. Res. 73,
571-584.

Reynolds, S. E., M. Brook, and M. F. Gourley (1957): Thunderstorm
charge separation. J. Meteorol. 14, 426-436.

Uzu, Y., and I. Sano (1970): Effect of the dissolved electrolytes
on the freezing of water - with particular reference to the
freezing potential. J. Meteorol. Soc. Japan 48, 255-257.

Workman, E. J. (1969): Atmospheric electrical effects resulting
from the collision of supercooled water drops and hail. Physics
of Ice. (Proc. Int. Symp. on Physics of Ice, Munich, Germany,
Sept. 9-14, 1968), pp. 594-602. Plenum Press, New York.

Workman, E. J., and S. E. Reynolds (1950): Electrical phenomena
occurring during freezing of dilute aqueous solutions and their

possible relationship to thunderstorm electricity. Phys. Rev.
78, 254-259.

EFFECT OF SOLUTE ON ICE-SOLUTION INTERFACIAL FREE ENERGY;

CALCULATION FROM MEASURED HOMOGENEOUS NUCLEATION TEMPERATURES

Don H. Rasmussen and Alan P. MacKenzie

Cryobiology Research Institute

RFD 5, Madison, Wisconsin 53704

INTRODUCTION

Among the forms of the water polymer interface of interest to low temperature scientists one distinguishes in particular the common surfaces separating ice from solid polymer and ice from aqueous polymer solution. The ice - solid polymer interface has been the subject of considerable attention; it was, for example, discussed at the Sapporo Conference on the Physics of Snow and Ice in 1966 /1/ and was the subject of a symposium chaired by Dr. H. H. G. Jellinek in 1967 /2/. Studies on the ice - aqueous solution interface have, in contrast, been limited largely to the interpretation of growth rate measurements in dilute salt solutions /3/, where the physical characteristics of the solution dominate the contribution from the surface effects.

In as much as the formation, growth, and recrystallization of ice in aqueous solutions is, in part, controlled by the physical and thermodynamic characteristics of the ice - solution interface, it is our purpose here to report studies on the aforementioned processes and to discuss, in particular, the effect of the solute on the interfacial free energy, the latter being a most important aspect. Because it is impossible to measure directly the free energy of a solid - liquid interface (the "surface tension"), the usual means of determination employ either (i) the measurement of equilibrium contact angles and grain boundary groove angles /4/, (ii) the perturbation of a smooth interface and observation of the growth or decay of the perturbation /5/ or, (iii) the measurement of the homogeneous nucleation temperature of the material /6/. Each of these methods involves theoretical assumptions as to the effects of surface energy on the process or the measurement. Within the validity

of the theoretical assumptions, a value for the interfacial free
energy can be calculated from the experimental results. Studies
of the interfacial free energy, γ , between ice and pure water by
the methods just mentioned are numerous and permit an interesting
comparison. Derived values for γ , all clearly of the same order
of magnitude, are presented in Table I. Where different investigators employed values for the homogeneous nucleation temperature
/6/ differing by no more than one or two degrees, slightly different reasoning resulted in the variation in the values for the interfacial free energy listed in the table.

Table I. Ice—Melt Interfacial Energy

Investigator	Method	Applicability	($ergs/cm^2$)
Dufour and Defay /7/	nucleation	average over all faces; T $\sim$ −40°C.	19.7
Fletcher /6/	nucleation	average over all faces; T $\sim$ −40°C.	22.0
Hardy and Coriell /5/	morphological stability	planes parallel to c-axis; T $\sim$ 0°C.	22.0
Hillig /8/	single crystal growth	edge energy of a step on basal plane; T $\sim$ 0°C.	6.4
Ketcham and Hobbs /4/	contact angle measurement	average over all faces; T $\sim$ −40°C.	33.0
Kotler and Tarshis /9/	dendritic growth	planes parallel to c-axis; T $\sim$ 0°C.	20.0
Schaefer /10/	nucleation	average over all faces; T $\sim$ −40°C.	32.1

We have applied the method of measurement of the homogeneous
nucleation temperature to determine the dependence of the surface
free energy on the solute concentration. We found we could determine the form of the concentration dependence without reference to
particular theories but not the numerical values of γ. While we
will indicate the shortcomings of the use of nucleation theory in
the derivation of γ, we will also try to demonstrate the value of
this experimental approach in the separation of the effects of different solutes on the stability of the ice - solution interface.

THEORY

To measure the temperature, T_h, at which a liquid cooled below its equilibrium melting point, T_E, will be spontaneously crystallized by the formation and growth of a nucleus of the solid phase, catalysts promoting nucleation must be removed from contact with the liquid. Frequently, liquids have been suspended in or on other inert liquids in attempts to prevent container walls from acting as sources of catalytic activity. Generally, however, pure liquids in bulk are still found to contain finite numbers of submicroscopic particles capable of acting as heterogeneous nuclei. Vonnegut /11/ and Turnbull and Vonnegut /12/ discovered that, in subdividing the bulk liquid into very small droplets such that the number of droplets far exceeded the number of particles, most of the droplets were free from sources of hetergeneous nucleation; that is, they would nucleate homogeneously by the spontaneous formation of solid phase. We reasoned, therefore, that if an aqueous solution was emulsified in an inert, insoluble carrier fluid to yield very fine drops of fairly uniform size (from 1 to 5 um. in diameter) where the emulsifying agent did not form a catalytic surface on the droplets, such an emulsion would cool until ice nuclei appeared spontaneously in the interior of the droplets.

The theory originally formulated by Volmer and Weber /13/ and Becker and Doring /14/ for nucleation of liquid droplets from the supersaturated vapor was modified by Turnbull /15/ for application to nucleation of crystals from the liquid state. The following presentation of nucleation theory necessary for evaluation of γ is based on that given by Uhlmann and Chalmers /16/, for the formation of spherical, isotropic nuclei.

Very small clusters appear in the melt only below the melting point and grow by a bimolecular reaction process. Since the surface to volume ratio is larger, the smaller the particle, the total surface energy constitutes a barrier to the growth of the cluster; that is, up to some critical size, the surface energy, which depends on r^2, where r is the radius of the cluster, increases with size faster than the volume energy, which depends on r^3, decreases. The sum of the free energy changes passes through a maximum at a critical cluster size, r^*. When an additional molecule is incorporated in the critical cluster, and r increases beyond r^*, a reduction in the total free energy is initiated. The cluster is then considered a nucleus and the process of crystal growth begins.

Mathematically, the free energy of formation, ΔG, of a cluster of radius r is given by:

$$\Delta G = (4/3)\Pi r^3 \Delta G_v + 4\Pi r^2 \gamma, \tag{1}$$

where ΔG_v is the volume free energy reduction in transferring water to ice. Equation (1), in which ΔG is expressed in terms of r, is depicted in component form in Figure 1.

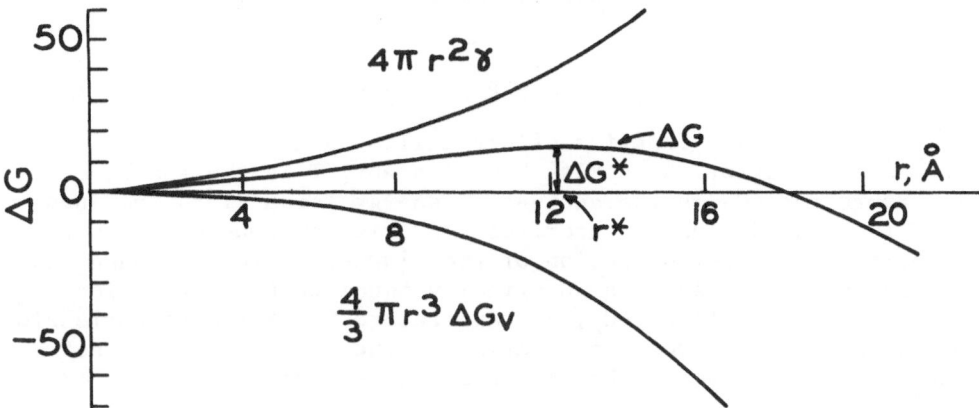

Figure 1. Free energy of formation, ΔG, of a cluster of radius r.

ΔG_v is given by:

$$\Delta G_v = - \Delta H_f \, (T/T_m) \, (\Delta T/T_m), \qquad (2)$$

where ΔH_f is the heat of fusion, T is the temperature of crystallization, T_m is the melting point and ΔT is the supercooling, $T_m - T$. Maximizing ΔG with respect to r we obtain the following relationship between r*, γ and ΔG_v:

$$r^* = - 2 \, \gamma / \Delta G_v, \qquad (3)$$

and the following expression for ΔG*:

$$\Delta G^* = 16\Pi \, \gamma^3 / \, 3\Delta G_v^2. \qquad (4)$$

Equations (3) and (4) are based on the assumption that thermodynamics can be applied to microscopic systems and, therefore, that ΔG_v and γ do not depend on the radius of the cluster, i.e. that they retain their macroscopic values. If it were experimentally possible to determine ΔG* at a temperature T (i.e., at a known value for ΔG_v) γ could be calculated from equation (4) and r* could be calculated from equation (3).

To evaluate ΔG*, the kinetics of homogeneous nucleation must be developed. The rate at which nuclei appear per unit volume per unit time is assumed to be given by:

$$J = A^* \omega \, N_i^* \qquad (5)$$

where A* is the surface area of the critical cluster, ω is the

frequency per unit area with which molecules of the right type jump
into the surface of the cluster, and N_i^* is the number of clusters
of critical size per unit volume. ω can be approximated for condensed systems by:

$$\omega = X_1 kT/A^*h \ \exp \ (- \Delta G_d/kT) \tag{6}$$

where X_1 is the mole fraction of crystallizing component, k is
Boltzmann's constant, h is Planck's constant, T is the absolute
temperature and ΔG_d is the activation energy for a diffusive jump
across the liquid cluster interface. ΔG_d is taken as equal to the
activation energy for diffusion of the slowest moving component in
the solution and is assumed numerically equal to the activation
energy for viscous flow, ΔG_m, at temperature T. For most nucleation
problems of interest where clustering in the liquid phase is not
expected, N_i^* is given by Boltzmann's statistics:

$$N_i^* = N_o \ \exp \ (- \Delta G^*/kT), \tag{7}$$

where N_o is the number of monomers present. When clustering is
prevalent in the liquid phase below T_m, Boltzmann's statistics do
not give the proper concentration for N_i^* and a much more complex
expression, for example, that used by Sundquist and Oriani /17/, is
required. In a structured liquid like water, one might therefore
question the use of equation (7). However, there being no other
simple measure of N_i^*, the Boltzmann expression is used, as it has
been in the past, to evaluate γ from homogeneous nucleation data.
Thus:

$$J = X_1 N_o (kT/h) \ \exp \ [- \ (\Delta G_m + \Delta G^*)/kT] \tag{8}$$

Since J is exponentially dependent on the reciprocal of the square
of the supercooling, (ΔG^* being proportional to $1/\Delta T^2$), the rate J
will change by orders of magnitude with small changes in ΔT. When
J approaches a rate of 1 nucleus per droplet per second at a temperature, T_h, all the droplets will crystallize in a short time,
barring temperature inhomogeneities due to thermal interaction between droplets during nucleation and crystal growth. T_h will be
used in the following to designate the temperature at which nucleation is observed experimentally.

J can be estimated by the volume of the drops, the cooling rate
and the temperature spanned during nucleation. We will assume:

$$J \sim 1 \ \text{nucleus/droplet/ sec.} \sim 10^{10}/\text{cc./sec.} \tag{9}$$

ΔG_m can be evaluated from viscosity data and T_h measured, so that
ΔG^* can be calculated as follows:

$$\Delta G^* = kT \ [\ln K_v - \ln J] - \Delta G_m \tag{10}$$

where K_v represents the term $X_1 N_o (kT/h)$ from equation (8). γ can, therefore, be calculated from equation (4) and r* from equation (3).

MATERIALS AND METHODS

According to Turnbull /18/, emulsification of an aqueous solution, that is, subdivision of a macroscopic sample into micron sized drops, should confine the limited number of contaminant particles, capable of heterogeneously nucleating the entire bulk sample, to a very small fraction of the droplets. If such an emulsion were then placed in an apparatus for differential thermal analysis and cooled at a finite rate, the crystallization of water in most of the droplets should result in an evolution of the latent heat, yielding an exothermic peak in the thermogram, only when the emulsion cooled through the homogeneous nucleation temperature. For a particular solute and solute concentration the nucleation temperature should prove reproducible with different emulsion carrier fluids and surfactants.

For the purpose of this analysis of the homogeneous nucleation temperatures of water and aqueous solutions, we used doubly distilled deionized water, Baker analyzed reagent grade ethylene glycol, glucose, glycerol, urea, NaCl, NH_4F, General Aniline and Film polyvinylpyrrolidone (Plastone C, dialyzed), Dow polyethylene glycol E-9000, Mallinckrodt analytical grade heptane, Dow Corning 702 fluid (silicone oil), and Atlas Chemical Industries Span 65 (sorbitan tristearate, HLB 2.1) and Arlacel 161 (glyceryl monostearate, HLB 3.8).

Emulsions of the water or aqueous solution were prepared in both carrier liquids, heptane and silicone oil, and stabilized by one of the surfactants, Span 65 or Arlacel 161. Heptane and Span 65 (4% w/w in the heptane) were usually employed as carrier fluid and surfactant, respectively, since no differences in the thermograms were observed to be due to the choice of carrier medium. Emulsions were prepared to contain 50% v/v water or aqueous solution and 50% v/v carrier fluid. 10 ml. of emulsion were prepared each time. The emulsion size was controlled by (i) simple shaking in a 35 ml. Pyrex specimen vial until microscopic observation revealed a uniform emulsion (average droplet diameter determined from photomicrographs of the emulsion was approximately 100 μm.), (ii) refinement to 20 μm. diameter droplets by forced flow from a syringe with a No. 26 needle and, (iii) further refinement by several passes through packed 100 mesh stainless steel gauze. The procedure yielded a uniform emulsion of approximately 3.5 μm. diameter droplets with standard deviation of 1.5 μm., calculated from the measurement of the diameters of approximately 1,000 drops.

The apparatus for differential thermal analysis consisted of
(i) a brass block that could be cooled, by controlled aspiration of
cold nitrogen gas, at constant rates from 1 to 15 deg. C./min. and
heated electrically at constant rates from 1 to 21 deg./min., (ii)
sample tubes of 1.0 mm. bore and .25 mm. wall thickness made of
Pyrex glass, (iii) a thermocouple system made from 3 mil copper and
constantan wires which measured (a) absolute reference temperature,
and (b) the temperature differential between sample and reference
and, (iv) an amplifier-recorder system for plotting, on an X - Y
basis, the differential temperature as a function of absolute reference temperature. The thermocouple-recorder system was calibrated
at the melting points of water (0°C.), and chlorobenzene (-45.2°C.).
The reference tube contained Baker analyzed reagent grade benzoic
acid. The setting of the ice point (0°C.) was checked before each
run. Temperatures were measured always to ± .3°C. The sensitivity
of the recorder to changes in the differential temperature was enhanced with the aid of a Leeds and Northrup d.c. amplifier; the resulting sensitivity could be varied (up to 0.5 μV/in. of display)
with minimal noise, though, for first order transitions such as
crystallization or melting, a typical differential temperature expansion of 10 μV/in. was employed.

The freshly prepared emulsion was pipetted into the sample
tube and the sample thermocouple inserted. Sample and reference
tubes were then placed in the brass block and cooled, generally at
a rate of 2 deg. C. per minute, to yield the thermogram. Nucleation at the homogeneous nucleation temperature resulted in the evolution of the latent heat of crystallization and an exotherm in the
thermogram at T_h. The droplets, being of a rather uniform size,
nucleated within a very narrow range of temperatures to yield a
sharp exotherm readily detected by differential thermal analysis.

RESULTS

In a previous experimental determination of the homogeneous
nucleation temperature of ice in distilled water and aqueous solutions /19/, the samples were prepared according to the procedure
just described with the exception that they were not submitted to
the third consecutive refining process. The fraction of the drops
containing heterogeneous nuclei proved, in that instance, not to be
negligible. For reasons that will become apparent, the results of
the earlier measurements on distilled water and on 20% ethylene glycol are presented in Figure 2. Each thermogram exhibited two distinct exotherms during cooling, the first representative of the
crystallization of the droplets in the emulsion containing particles
causing heterogeneous nucleation at lesser supercoolings; the second peak, at the lower temperature, indicated the homogeneous nucleation of ice in the remaining droplets. The uneven appearance of
the exotherm denoting homogeneous nucleation was attributed to the

considerable variation in the size of the droplets. The refinement
to which the emulsions were subjected in this, the more recent study
served to confine the heterogeneous nuclei to a very small proportion of the droplets. Little or no heterogeneously nucleated crystallization was observed. At the same time, the droplet size was
much more nearly uniform and the peaks in the thermograms denoting
the crystallization of homogeneously nucleated ice were very much
smoother, and were limited to narrower temperature ranges (from 1.5
to 2 deg. C., generally). Figure 3 shows the type of thermogram
obtained with deionized distilled water and with 20% w/w glucose.
The single exotherm during cooling indicates a sharp onset temperature for the homogeneous nucleation of ice at -37°C. and a maximum
rate of crystallization at -38.3°C. In the case of 20% glucose the
onset temperature was -43.8°C. and the temperature at the peak was
-45.2°C. We have followed the practice of equating the measured
peak temperature with that of homogeneous nucleation /20/.

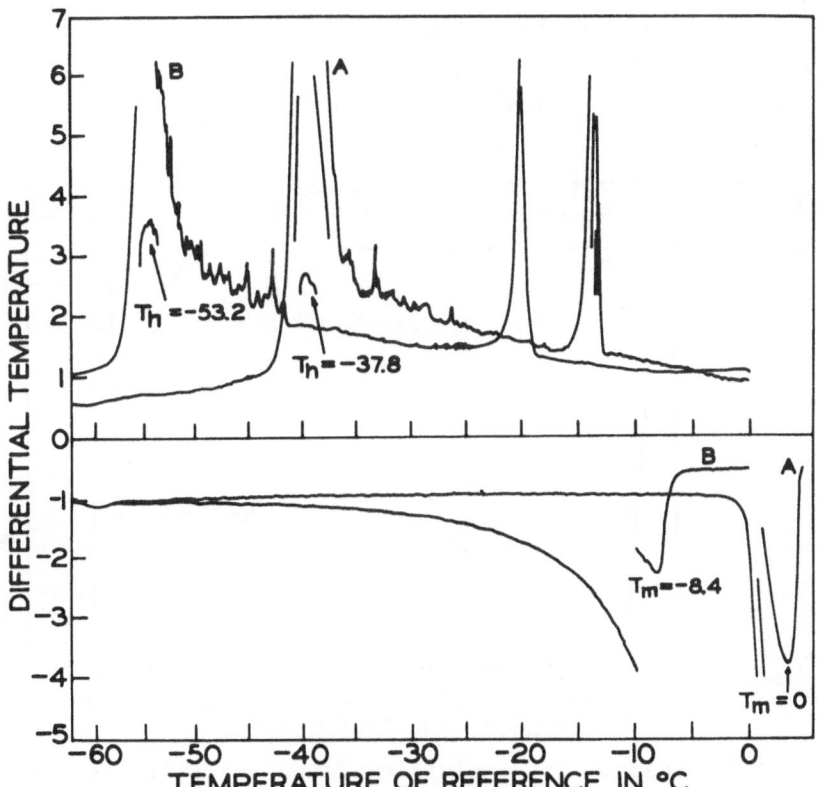

Figure 2. Thermograms for emulsified deionized distilled water, A,
and for emulsified 20% ethylene glycol, B. Upper curves: d.t.a. on
cooling, exotherm corresponding to homogeneous nucleation marked T_h.
Lower curves: d.t.a. on warming, endotherm for melting marked T_m.
Ordinate values are scale units only. (See reference 19).

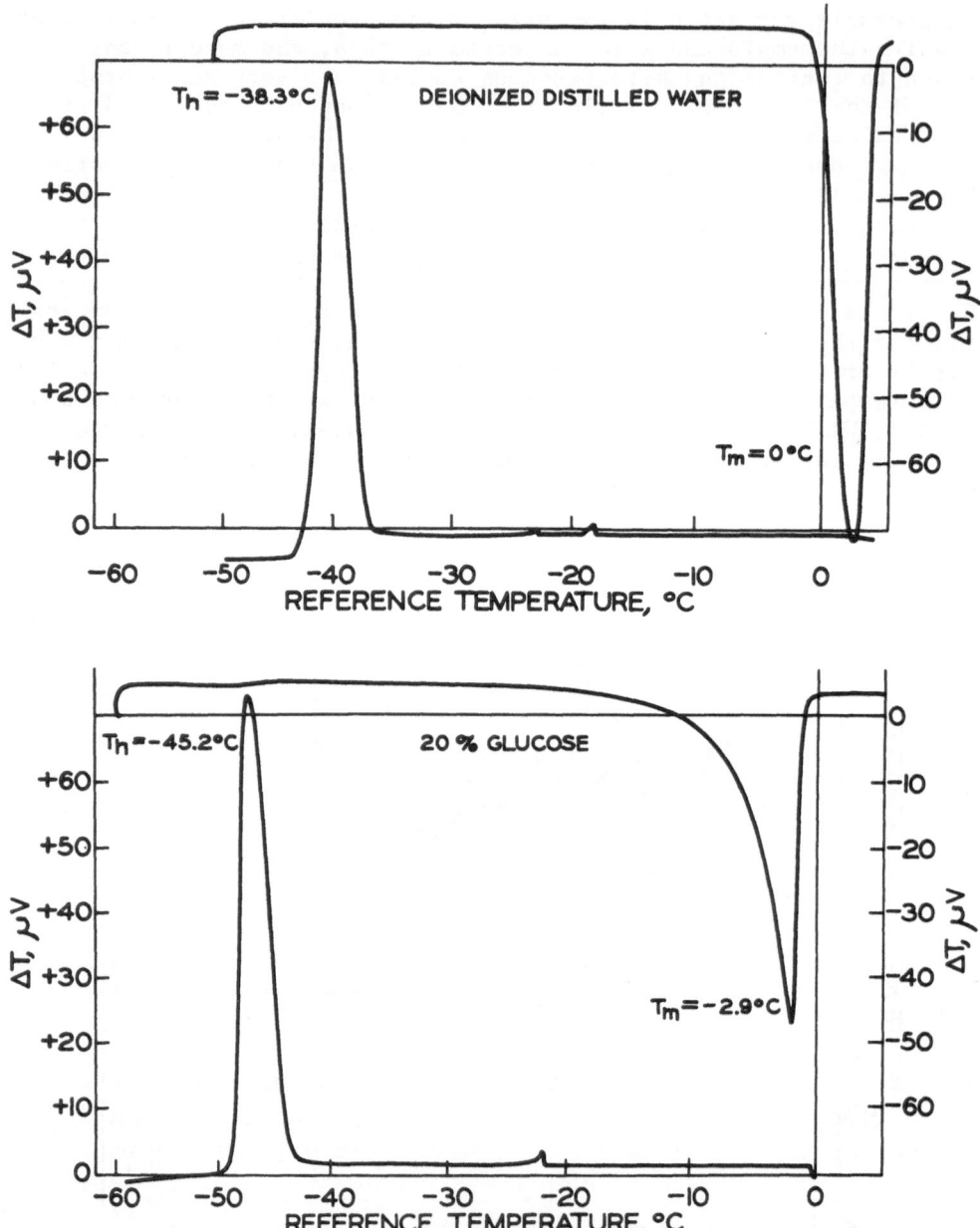

Figure 3.   Thermograms for emulsified deionized distilled water and for 20% glucose.   In each thermogram, the lower curve is d.t.a. on cooling, and the upper curve is d.t.a. on warming.   Note the clear onset of homogeneous nucleation on cooling; exotherm marked T_h.

To interpret the findings in terms of homogeneous nucleation theory, it was necessary to know the precise extent of the supercooling. Supplement thermograms were therefore obtained on warming to provide the melting points of the emulsified solutions (see again Figures 2 and 3). It was observed in every case that the measured melting point corresponded to the melting point determined on a bulk quantity (that is, on an unemulsified sample), of the same solution, within the instrumental accuracy of ± 0.3°C. The melting points of the solutions were, that is, affected neither by the carrier fluid nor by the surfactant, nor by the process of subdivision inherent in emulsification.

Experimentally determined homogeneous nucleation temperatures proved to be highly reproducible; the measured values were not altered when the samples were warmed 40 degrees above the melting point; nor were they affected by choice of carrier fluid and surfactant. Such observations support our contention that we were indeed observing homogeneous nucleation. We note that we obtained a T_h for water of -38.3°C., in excellent agreement with previously published values, (i) by direct observations on distilled water, (ii) from determinations on aqueous solutions, by extrapolation to 0 solute concentration. Measurements of the type described were made on a variety of solutes, at a number of concentrations. Selected data are presented in Table 2.

Table 2. Experimental Data

Solute	wt.%	T_m	T_h	Solute	wt.%	T_m	T_h
Distilled water	0.00	-38.3		NaCl	15	-11.1	-57.6
					20	-17.0	-68.2
EG	10	-3.0	-44.2				
	15	-5.8	-49.8	NH_4F	5	- 4.2	-46.1
	20	-8.3	-55.1		10	- 8.4	-54.0
	25	-11.4	-61.6		15	-13.2	-63.7
	30	-15.2	-69.3		20	-18.2	-76.4
	35	-19.2	-78.5				
	40	-24	-88.0	Glucose	10	- 1.2	-41.3
					20	- 3.0	-45.2
Glycerol	20	- 5.1	-47.6		30	- 5.1	-50.2
	30	-10.2	-57.5		40	- 7.7	-56.5
	40	-15.8	-69.0				
				Carbowax	20	- 1.1	-44.2
Urea	5	- 2.1	-41.5	E-9000	30	- 2.7	-53.0
	10	- 3.6	-44.5		40	- 5.8	-67.1
	20	- 7.4	-51.0				
	30	-11.2	-57.9	Polyvinyl	20	- .4	-39.6
				Pyrroli	30	- .9	-40.9
NaCl	5	- 3.3	-43.6	done	35	- 1.6	-43.5
	10	- 6.8	-50.0		40	- 2.8	-48.4

To test the effect of different cooling rates on the measured
homogeneous nucleation temperature, emulsions of distilled water
and of 30% ethylene glycol were cooled at rates from 1 to 13 deg./
min. For the distilled water a measured T_h of -38.7°C. at a cooling rate of 13 deg./min. compared with -38.3°C. at 1 deg./min. The
0.4 deg. difference falls within the experimental error. For the
30% EG the measured T_h ranged from -69.5°C. at 11 deg./min. to
-69°C. at 1 deg./min. While the 0.5 deg. difference is again within the accuracy of the method, there is a suggestion of a limited
additional depression with increased cooling rate. In this respect, it might have been more important to determine the T_h at
lower cooling rates, e.g., at 0.1 deg./min. /21/ which, however,
we were unable to achieve in our apparatus.

DISCUSSION

The surface energy and the critical radius of the nucleus can
be calculated from the measured nucleation temperature and the supercooling provided reasonable estimates are made for the various
terms in the theoretical rate expression (equation 8). The results
of the calculation of ΔG_m, the activation energy for viscous flow,
in the case of aqueous solutions of ethylene glycol and polyvinylpyrrolidone, together with evaluations of γ and r*, are presented
in Table 3. The only quantity not determined by the emulsion drop
size or from the observed thermograms is the activation energy for
viscous flow at T_h. Since measurements of the viscosity of a supercooled solution are practically impossible in the neighborhood of
T_h, we estimated the activation energy for viscous flow, extrapo-
lating the values from published viscosity data taken from above
the melting point to the measured glass transition temperature,

Table 3. Calculations of γ and r*

Conc.	(lnK-lnJ)	$\Delta G_m/kT_h$	$\Delta G^*/kT$	γ (ergs/cm^2.)	r* ($\overset{\circ}{A}$)
Solute: Ethylene glycol					
0	58.8	15	43.8	22.4	12.3
10	58.7	19.8	38.9	22.3	11.5
20	58.6	26.9	31.7	21.6	10.3
30	58.6	36.6	22.0	19.8	8.6
40	58.5	47.6	10.9	16.0	6.5
Solute: PVP					
0	58.8	15	43.0	22.4	12.3
20	58.8	33.4	25.4	19.0	10.1
30	58.8	40.9	17.9	17.0	9.0
40	58.8	53.8	5.0	12.0	6.0

where viscosities are assumed equal to 10^{13} poise. Values for the
viscosity for the ethylene glycol – water system were taken from
Curme and Johnston /22/; glass transition temperatures were measured
in a previous study /23/. The viscosity for PVP was taken from the
General Aniline and Film Co. brochures, "PVP." /24/. The glass transitions used in the extrapolation of the PVP viscosity/temperature
curves are taken from an extrapolation of the T_g versus water content values presented in our companion paper on PVP presented at
this symposium /25/.

The results of the calculations of γ and r*, presented in
Table 3, indicate that both ethylene glycol and polyvinylpyrrolidone
decrease the surface energy and radius of the critical nucleus at
T_h. It is surprising that the soluble polymer should decrease the
free energy of the ice solution interface more than ethylene glycol
which has the greater capability to hydrogen bond with the ice surface. The reductions in the surface free energy, γ, with increased
solute concentration suggest adsorption of both solutes by the nucleus. But the critical radius of the nucleus, r*, in, for example,
40% PVP is no more than 6 Å at T_h. The number of water molecules
in such a nucleus cannot, therefore, exceed about 30; that is,
the molecular weight of the nucleus cannot exceed a figure of 500
to 600. When the size of the nucleus is compared with the size of
the polymer (ca. 30,000 mol. wt.) it is seen that it is perhaps better to visualize the sorption of the nucleus to the polymer.

The values of γ and r* calculated for ice in solutions of EG
and PVP depend on estimated values of ΔG_m. $\Delta G_m/kT_h$ varies from
values small with respect to $\Delta G*/kT_h$ at low solute concentrations
to values large with respect to $\Delta G*/kT_h$ at higher solute concentrations. Since $\Delta G*/kT_h$ represents the difference between an approximately constant quantity ($\ln K_v - \ln J$) and $\Delta G_m/kT_h$, any error in
$\Delta G_m/kT_h$ is an error in $\Delta G*/kT_h$ and therefore also in the value of
γ calculated from it. One should note that, in this respect that,
in the case of a polymer, the activation energy for viscous flow may
not equal the activation energy for diffusion of the solvent water,
or for that matter, that for the localized diffusive movement of
parts of the polymer necessary to permit clustering of the water molecules. The molecular wt. of the PVP (30,000) is sufficiently large
that the bulk properties of its concentrated solutions will be influenced by the spatial density of the polymer and by possible en-
tanglement coupling.

Further examination of the experimental data casts some doubt
on the methods employed to calculate γ. The depression of the homogeneous nucleation temperature is greater than the depression of the
melting point for every solute investigated. This is clearly demonstrated in Figure 4 which consists of a "phase diagram" on which
the temperatures of homogeneous nucleation have been plotted against
weight fraction solute. The melting point curves, originating at

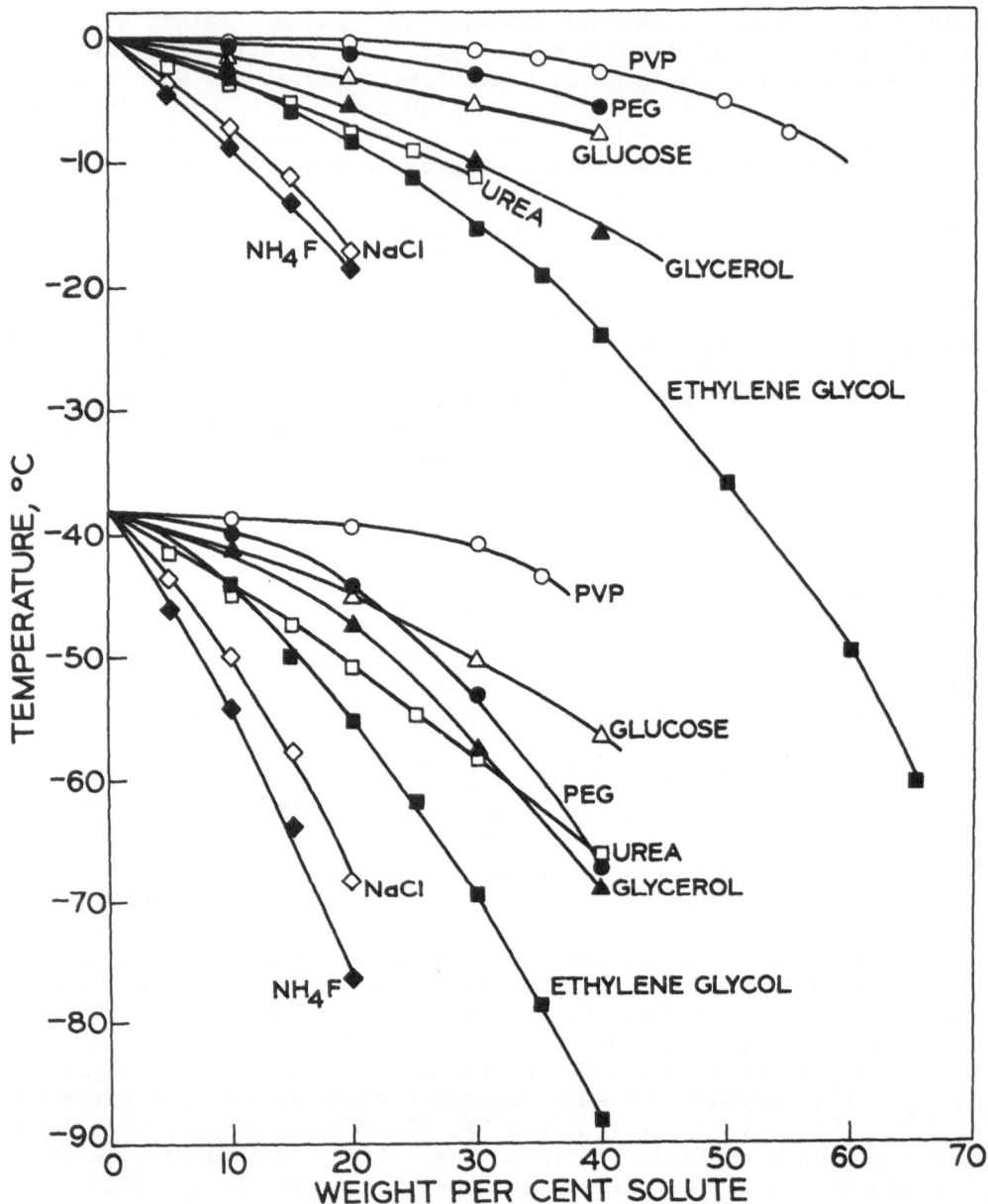

Figure 4.   'Phase diagram' containing melting point depression
curves and homogeneous nucleation temperature depression curves for
the solutions listed in Table 2.   Compare the shape of the homogeneous nucleation temperature depression curve with the corresponding melting point depression curve.

0°C., are seen to be clustered above the homogeneous nucleation temperature curves describing the spontaneous formation of ice on cooling, the latter each originating at -38.3°C. The characteristic of greatest importance is the general shape of the homogeneous nucleation temperature curve, which is nearly the same as the shape of the melting point depression curve for every solute.

Because the melting point depression curve and the homogeneous nucleation temperature curve have very nearly the same shape, though they occur at different temperatures, we have compated the nucleation temperature with the corresponding melting point. The results of this comparison for all the solutes studied is presented in Figure 5. The linearity of the plots of T_h versus T_m in this figure

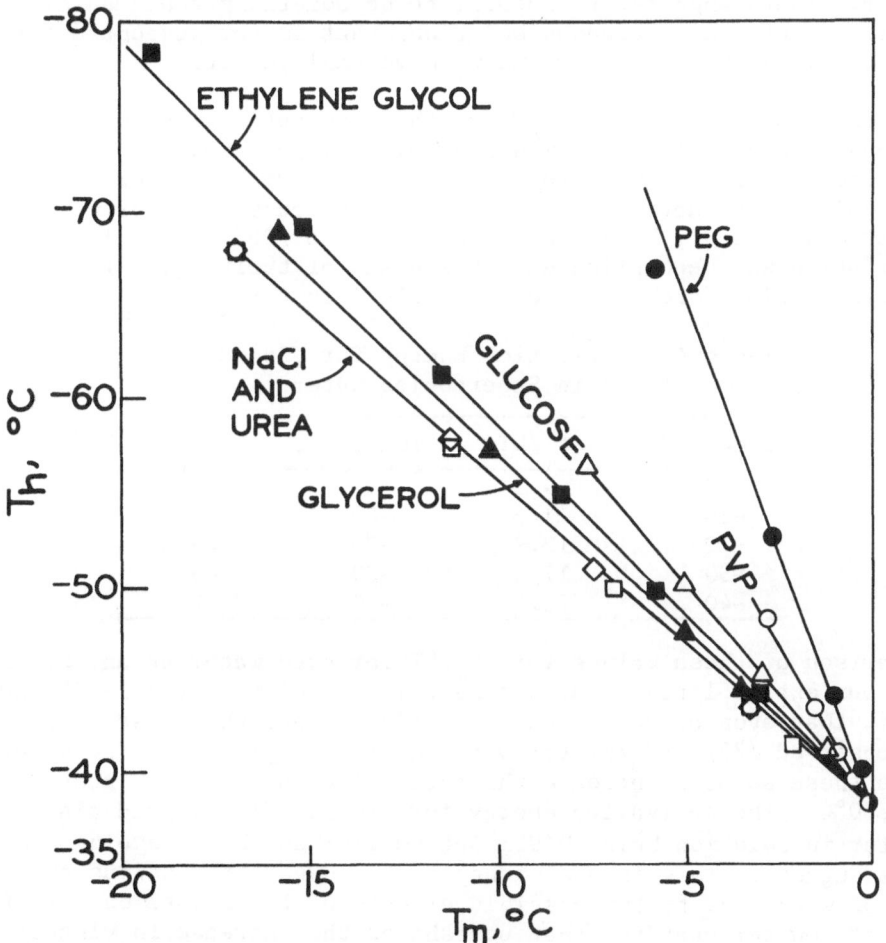

Figure 5. Homogeneous nucleation temperature, T_h, versus melting temperature, T_m, for the solutions listed in Table 2.

is quite surprising since the melting point depression is due to reduction in water activity with increased solute concentration while
the depression in the homogeneous nucleation temperature results
from the reduction in water activity and from changes in the activation energy for diffusion, changes in the ice solution interfacial
free energy, and, possible, from other effects of ice-like structuring or destructuring at constant water activity.

According to the experimental design which required similar
emulsion droplet sizes and cooling rates, the rates of nucleation,
J, for different concentrations of solute should have been about
equal at the respective T_h's. Assuming this equality, and because
of the concentration and temperature dependence of $\Delta G_m/kT_h$, we were
unable to deduce an expression that related T_h to T_m in a linear
manner. There appears, therefore, to be something fundamental to the
nucleation of ice in aqueous solutions that is not accounted for by
the theory in the form in which we have employed it.

In an attempt to determine whether the activation energies for
viscous flow calculated from graphical extrapolation of the viscosity, are of a reasonable order of magnitude, we have, for comparison,
extrapolated the activation energies for viscous flow for pure water
reported by Korson, Drost-Hansen and Millero /26/. This latter extrapolation was accomplished on the basis of their eq. 2 /26/. The
resultant values are presented in Table 4 in terms of $\Delta G_m/kT$.

Table 4. Activation Energy for Viscous Flow
in Supercooled Water

Temp. °C.	$\Delta G_m/kT$	Temp. °C.	$\Delta G_m/kT$
0	9.5	−50	28.6
−10	11.3	−60	37.3
−20	13.9	−70	48.8
−30	17.4	−80	63.9
−40	22.2	−90	82.8

Comparison of these values for $\Delta G_m/kT$ for pure water at any given T_h
with our extrapolated values of $\Delta G_m/kT_h$ in solution (Table 3) indicates, that even at solute concentrations where the glass transition
was observed /23/, $\Delta G_m/kT$ for water exceeds $\Delta G_m/kT_h$ for the solution.
Since these solutes increase the activation energy for viscous flow
above 0°C., the activation energy for viscous flow should also be
greater in solution below 0°C., not smaller as the foregoing com-
ments suggest. This inconsistency may be due to the unique structure of water and to the possible effects of the solute on the structure at low temperatures that overshadow the increase in viscosity
that the solute imparts at room temperature.

Further discussion of surface energies calculated from homogeneous nucleation theory, based on such inconsistent values for the activation energy for viscous flow does not appear to be justified. We will, therefore, limit ourselves to considerations of the surface energy per unit volume of the nucleus, ($2\gamma/r*$), this parameter being obtained directly from the Kelvin equation for the solubility of finely divided solids. This equation, given earlier (eq. 3), is the basis for the energetics of the nucleation process:

$$2 \quad \gamma/r* = -\Delta G_v. \qquad (3)$$

The driving free energy, ΔG_v, can be given by either of the following formulae,

$$-\Delta G_v = \Delta H \, T \Delta T \, / \, T_m^2 \qquad (2)$$

as given in the theory section, or by

$$\Delta G_v = (RT/ \, \Omega) \, \ln \, (a_{r*}/a_\infty) \qquad (11)$$

where R is the gas constant, T is the absolute temperature, a_{r*} is the activity of water in ice crystals with radius r*, a_∞ is the activity of water in a solution in equilibrium with ice at T_h, and Ω is the volume per molecule in the ice nucleus. The use of eq. 11 for ΔG_v requires a system exhibiting measureable melting points lower than the highest measured nucleation temperatures; the ethylene glycol – water system provides the necessary data. If the temperature of homogeneous nucleation determined experimentally is based on the concept that nuclei are at their metastable melting points, then ($2\gamma/r*$) can be calculated either from (2) or (11). It should be noted that the viscosity does not appear in either expression and if the viscosity has lowered the rate of nucleation, the experimentally observed supercooling will be increased over its equilibrium value and ($2\gamma/r*$) calculated from eqs. 2 & 3 will not equal ($2\gamma/r*$) calculated from eqs. 3 & 11. The calculated values for ($2\gamma/r*$) for the ethylene glycol – water system at T_h are presented in Table 5. The activity of water at T_h in equilibrium with a cluster of radius

Table 5.   Calculated Values for ($2\gamma/r*$) at T_h in Aqueous Solutions of Ethylene Glycol

Temperature	a_{r*}	a_∞	($2\gamma/r*$) x 10^7 eqs. 3 & 11	($2\gamma/r*$) x 10^7 eqs. 2 & 3
-40°C.	.990	.678	36.7 ergs/cc	37.2 ergs/cc
-45	.968	.648	38.5	39.2
-50	.950	.619	40.2	41.0
-55	.933	.589	42.2	42.8
-60	.917	.560	44.1	44.6

r*, a_{r*}, has been put equal to its mole fraction for lack of a better estimate. The activity of macroscopic ice, a_{∞}, is equal to the ratio of the vapor pressure of ice over the vapor pressure of supercooled water at T_h. The other quantities have been previously defined. The values of $(2\gamma/r*)$ in Table 5 are almost identical and the difference between them shows no trend as a function of temperature. This suggests that the Kelvin equation applies to the critical nucleus and that the viscosity has not drastically reduced the nucleation rate assumed in the calculation of γ in Table 3. It is important to note that, with reference to the water in the clusters the activity coefficient of the water external to the cluster is equal to 1 at T_h; that is, the solution of clusters of critical size exhibits ideal behavior. This may be one of the more important conclusions to be drawn from our results.

The surface energy per unit volume of the nucleus has been calculated for all the solutes listed in Table 2 and is presented as a function of T_h in Figure 6. A solute that increases the surface

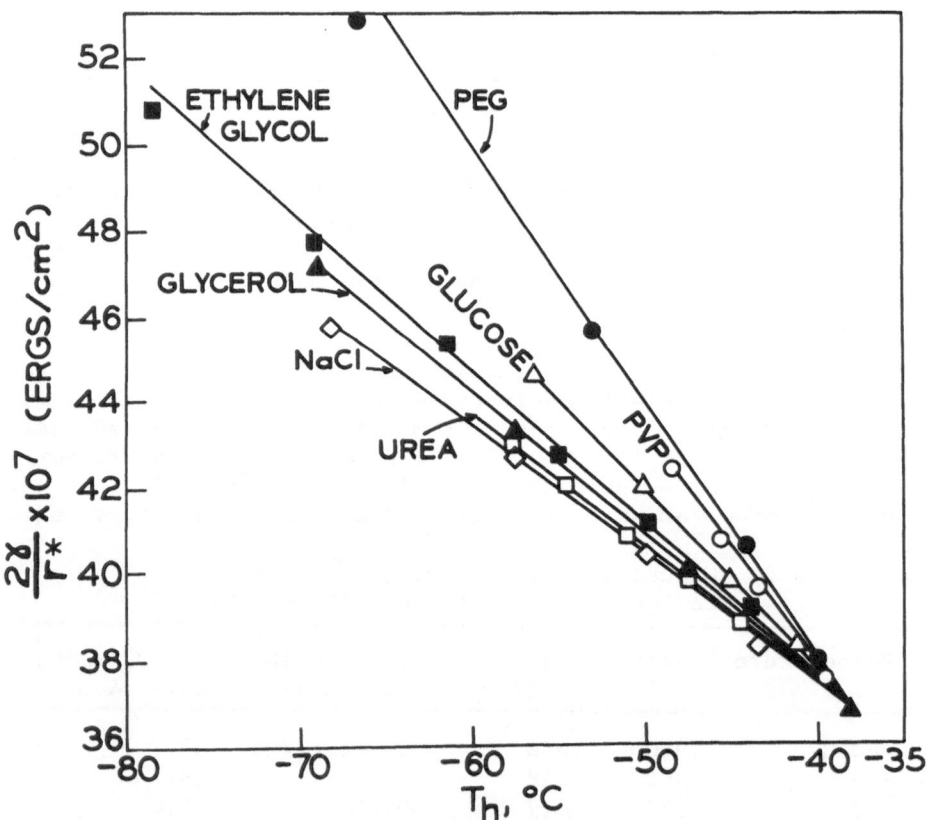

Figure 6. Surface energy per unit volume of nucleus, $2/r*$, versus T_h for the solutions listed in Table 2.

energy per unit volume of the nucleus can be expected either to increase the surface energy, γ, or to decrease the radius of the nucleus, $r*$. Intuitively, increased values for γ imply reduced values for $r*$, and vice versa. By comparison of the effect of the different solutes on $(2\gamma/r*)$ the expected form of the concentration dependence of the surface energy can be deduced. The calculated results presented in Figure 6 correlate well with the experimental results presented in Figure 5. The soluble polymers, PVP and polyethyleneglycol (PEG) depress the homogeneous nucleation temperature, T_h, 3.5 and 5.2 times more than they depress the melting point, respectively, and also increase $(2\gamma/r*)$ more at T_h than do the low molecular weight solutes. This is to be expected because of the dependence of $(2\gamma/r*)$ on ΔT_h (from equation 3). Since an increase in $(2\gamma/r*)$ can be accomplished more easily by a slight increase in γ and a slight decrease in $r*$, and all of the solutes studied increase $(2\gamma/r*)$, we might conclude that γ is increased by all the solutes. γ need not increase, however, for is $r*$ is reduced significantly then γ may stay about constant or even decrease. It is unlikely that in the case of the polymer solutes where $(2\gamma/r*)$ is increased the most with concentration that γ could be decreased unless a very minute nucleus, of the order of 20 to 50 water molecules could be stabilized (this is an order of magnitude smaller than the size of the nucleus calculated for pure water /6/). We conclude then that the calculation of γ given in Table 3, based as it is on an accepted rate expression, is essentially fortuitous and, that homogeneous nucleation theory should be modified, in the case of the nucleation of ice from solution, the better to describe the observed experimental results.

Resolution of the difficulties encountered in interpretation of experimental results may depend on further research into the structure of water and aqueous solutions. Perhaps, if ice-like structures were present in supercooled water /27/, the effect of solute on the ice-like structures would explain the effect of solute on the homogeneous nucleation temperature; γ could then be calculated from an appropriate modification of nucleation theory /28/. The effect of solute on the ice-like structures could also explain the problems encountered in the evaluation of the activation energy for diffusion at subzero temperatures. Until such an analysis is completed, we conclude that it is most likely that polymers such as PVP and PEG increase the ice solution interface free energy while the small molecules and salts we studied most probably reduce the surface free energy but the reservations expressed in the course of this discussion should be kept in mind.

ACKNOWLEDGEMENT

This work was supported by Grants from the ONR (NONR-3700) and from the NIH (GM-15143).

REFERENCES

/1/ Oura, Hirobumi, Ed. Physics of Snow and Ice, (The Institute
     of Low Temperature Science, Hokkaido University) (1967).

/2/ Jellinek, H.H.G., Chmn, "Ice Symposium," J. Coll. Interface
     Sci. 25, 131-294 (1967).

/3/ Pruppacher, H. R., J. Coll. Interface Sci. 25, 285-294 (1967).

/4/ Hobbs, P. V. and Ketcham, W. H., In Physics of Ice, Eds. Riehl,
     N., Bullemer, B. and Engelhardt, H. (Plenum Press, New York)
     95-112 (1969).

/5/ Hardy, S. C. and Coriell, S. R., J. Crystal Growth 7, 147-154
     (1970).

/6/ Fletcher, N. H., The Chemical Physics of Ice, (Cambridge Uni-
     versity Press, London) (1970).

/7/ Dufour, L. and Defay, R., Thermodynamics of Clouds, (Academic
     Press, New York) (1963).

/8/ Hillig, W. B., in Growth and Perfection of Crystals, Eds. Dore-
     mus, R. H., Roberts, B. W., and Turnbull, D. (John Wiley and
     Sons, New York) 350-360 (1958).

/9/ Kotler, G. and Tarshis, L. A., J. Crystal Growth 3-4, 603-610
     (1968).

/10/ Schaefer, V. J., Ind. Eng. Chem. 44, 1300-1304 (1952).

/11/ Vonnegut, B., J. Colloid Sci. 3, 563-569 (1948).

/12/ Turnbull, D. and Vonnegut, B., Ind. Eng. Chem. 44, 1292-1297
     (1952).

/13/ Volmer, M. and Weber, A., Z. Phys. Chem, 119, 277-301 (1926).

/14/ Becker, R. and Doring, W., Ann. Phys. 24, 719-752 (1935).

/15/ Turnbull, D. and Fisher, J. C., J. Chem, Phys. 17, 71-73 (1949).

/16/ Uhlmann, D. R. and Chalmers, B., in Nucleation Phenomena,
     Michaels, Alan S. Chmn, (American Chemical Society Publica-
     tions, Washington, D. C.) 1-15 (1966).

/17/ Sundquist, B. E. and Oriani, R. A., J. Chem. Phys., 36, 10,
     2604-2615 (1962).

/18/   Turnbull, D., J. Appl. Phys, 21, 1022-1028 (1950).

/19/   Rasmussen, D. and Luyet, B., Biodynamica 11, 33-44 (1970).

/20/   Kuhns, I. E., J. Atmos. Sciences, 25, 878-880 (1968).

/21/   Cormia, R. L., Price, F. P. and Turnbull, D., J. Chem. Phys.,
       37, 6, 1333-1340 (1962).

/22/   Curme, G. O. and Johnston, F., Glycols, 57, (ACS Monograph,
       Reinhold Publishing Corp., New York) (1953).

/23/   Rasmussen, D. H. and MacKenzie, A. P., J. Phys. Chem. 75,
       967-973, (1971).

/24/   General Aniline and Film Corp., New York, PVP (1964).

/25/   MacKenzie, A. P. and Rasmussen, D. H., this volume.

/26/   Korson, L., Drost-Hansen, W., and Millero, F. J., J. Phys.
       Chem. 73, 34-39 (1969).

/27/   Gilra, N. K. and Dass, N., J. Physical Soc. Japan, 24, 4, 910-
       912 (1968).

/28/   Frenkel, J., Kinetic Theory of Liquids, 382-400 (Oxford at
       the Clerendon Press) (1946).

INTERACTIONS IN THE WATER-POLYVINYLPYRROLIDONE SYSTEM

AT LOW TEMPERATURES

Alan P. MacKenzie and Don H. Rasmussen

Cryobiology Research Institute

RFD 5, Madison, Wisconsin   53704

INTRODUCTION

Great interest has attached to the hydration of macromolecular species and to the state of water molecules in hydrophilic polymer systems in particular. Bull /1/ and Hoover and Mellon /2/ reported the sorption of water by cellulose, cellulose acetate, nylon, and numerous proteins, applying equations developed originally to describe the binding of monatomic gases and small molecules to free surfaces. Bull determined the method of Brunauer, Emmett and Teller /3/ to suffice at 25 and 40°C. from 0.05 to 0.50 water activity (denoted hereafter by the symbol 'a_w'). Hoover and Mellon applied Bradley's isotherm /4,5/ very successfully throughout the range of a_w's (0.06 to 0.93) encompassed in their studies. Bull, and Hoover and Mellon each determined that water became attached to certain sites on the macromolecules, also that additional water molecules became attached to the first to form layer-like structures.

Each of the theoretical treatments has its merits. Bradley's equation clearly describes the extent of the water binding over the greater range of water activities — Ling /6/ has, in this respect, offered an excellent demonstration of the superiority of the Bradley treatment. Conversely, the B. E. T. analysis is especially useful inasmuch as it usually permits the derivation of constants to which a simple pictorial significance is readily attributed.

Recently, Jellinek, Luh and Nagarajan /7/ reported water sorption studies on several polymers in which only the respective side chains possessed the potential for specific interaction with the sorbate. From a thermodynamic analysis, Jellinek et al. determined water bound directly to polyacrylonitrile (PAN) and polymethylmethacrylate (PMM) molecules to be less strongly bound and less well ordered than the water sorbed in the form of succeeding layers. From a corresponding analysis of their measurements on polyvinylpyrrolidone (PVP) at -2, 7.5 and 10°C. these same authors concluded that, during resorption, water assumed ice-like structures adjacent to the polymer chains. A less substrate-specific interaction was, apparently, inferred in the system: water - PVP than in the systems: water - PAN and water - PMM.

More recently, largely to help determine the means by which PVP protects human red blood cells and certain cells in tissue culture from freezing injury /8,9,10,11/, and yeast cells from death during freeze-drying /12/, we undertook the determination of the sorption of water by this polymer at temperatures ranging from -40 to 22°C. To standardize, as far as possible, the thermal histories of the starting materials and to decrease, as much as we could, the times required to complete the sorption measurements, we prepared the samples from 20% aqueous PVP by freezing and freeze-drying. To prepare the material for desorption below 0°C., we employed a form of freeze-drying in which the frozen specimens were exposed to water vapor exerting vapor pressures just less than those of the respective specimens.

To distinguish glassy states from viscous states we determined the dependence of the glass transition temperature on the water content by differential thermal analysis. Consequently we were able to discuss desorption and resorption with reference to passage from viscous to glassy states and vice versa. We had already found that by freezing dilute aqueous PVP to -40°C., that is, by causing ice to form and aqueous PVP to dehydrate, we could reduce the polymer-rich phase to the glassy state, and effect the desorption, subsequent to the sublimation of ice at -40, entirely from a rigid matrix. At the higher temperatures, desorption to water contents lower than those obtained by freezing was required to cause the PVP to harden.

We will attempt to show that the concept of the fixed interface permits the description of the sorption data to 0.50 a_w, or thereabouts, and to higher activities to the extent that the water molecules are subject to dipole-dipole interactions with the PVP and, no less important, with each other. We will also discuss the likelihood that the values we have assumed the vapor pressure of liquid water to possess in the range: -40 to 0°C. are in greater error the lower the temperature. Lastly, we will examine the paradox inherent in the seemingly steady rise in the height of the desorption and the resorption isotherms with decreasing temperature, proposing an increasing ice-likeness in the water retained in the polymer phase in the presence of ice, the colder the system.

MATERIALS

Polyvinylpyrrolidone designated PVP K-30 was obtained from the General Aniline and Film Corporation. 100-g. quantities, dissolved in ca. 400-ml. volumes of distilled water, were freed from low molecular weight fractions by dialysis for 48 hr. at 2°C. The PVP was recovered in the form of a light powder by conventional laboratory freeze-drying.

METHODS

Isotherm Determination

At -40, -30, -20 and -10°C.   Each isotherm was constructed from a series of measurements on a single sample. Every sample was, that is, exposed, in succession, at constant temperature, to water vapor atmospheres of different water activities. Figure 1 illustrates, in principle, the system adopted to permit the separate regulation of the sample temperature and the water activity. It is

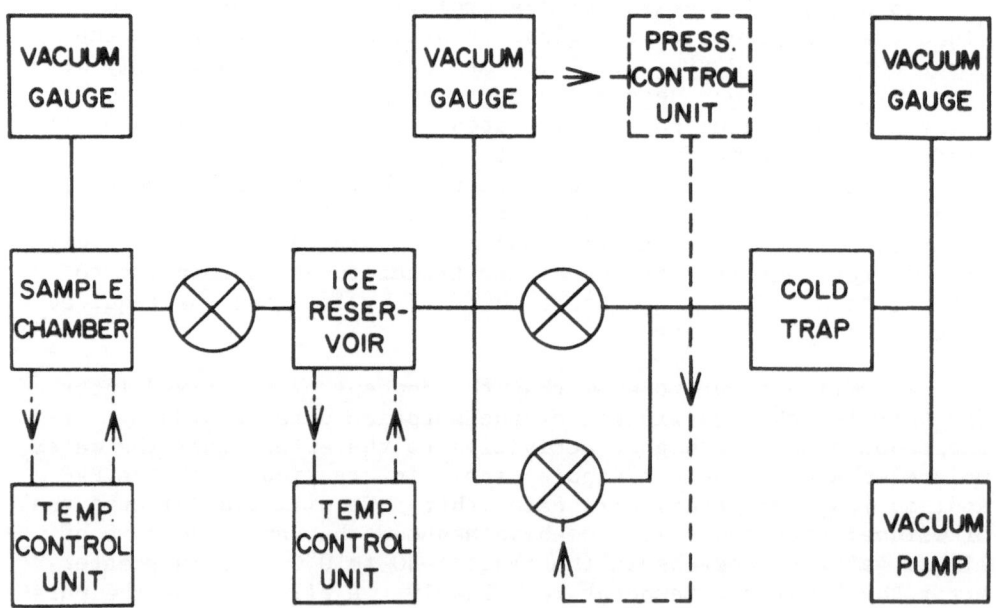

Figure 1. Schematic illustrating the means employed to control the water activity during the sub-zero sorption measurements. Samples kept at constant temperature were exposed to water vapor tending to equilibrium with ice maintained at another, lower temperature.

readily seen that the regulation of the water activity depended, in large part, on the control of the temperature of a second enclosure. The set-up we used is shown in detail in Figure 2.   The vacuum apparatus, part glass, part metal, incorporated a Cahn RH recording balance, and a variety of valves, gauges, traps, and pumps. Sample assemblies were suspended from the balance beam by means of a fine metal ribbon to which were attached, just above the sample, one or more radiation shields (discs made from thin aluminum foil).

    Sample chamber and ice reservoir temperatures were controlled each to ± 0.05°C., by partly electrical, partly pneumatic systems of the type described by Luyet and Rapatz /13/.   These and the sample temperature were recorded on a single chart to permit later calculation of the sample/ice reservoir temperature difference free from instrumental zero point errors. The temperature within the sample was measured with the aid of a thermal junction, the 3-mil leads from which were attached to the suspending ribbon and to the

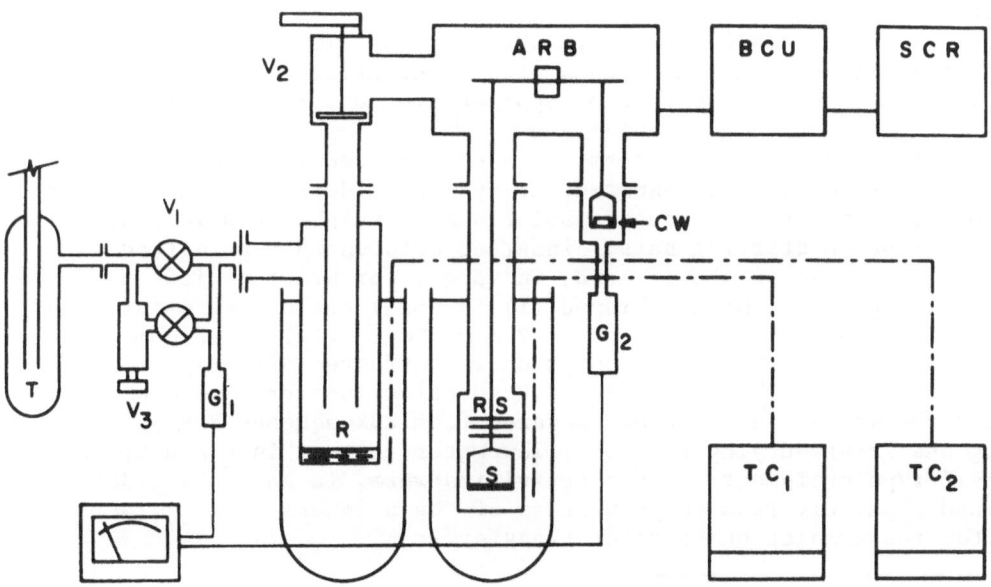

Figure 2. Diagram depicting the system constructed for the determination of the desorption and the resorption isotherms from single samples. ARB: Cahn RH automatic   recording balance; BCU: balance control unit; CW: counterweights; G_1, G_2: vacuum gauges; R: ice reservoir; RS: radiation shields; S: sample; SCR: X - t strip-chart recorder; T: refrigerated trap; TC_1, TC_2: Honeywell Air-O-Line temperature controllers connected to the sample chamber bath and the ice reservoir bath, respectively (stirrers not shown); V_1, V_2: wide-bore valves; V_3: adjustable needle valve.

balance beam and connected to fixed terminals on the balance chassis such that the operation of the balance was not affected /14/.

To prepare for a determination, 2 ml. of 20% aqueous PVP were frozen by gentle immersion in liquid nitrogen, transferred to the sample chamber, and allowed to warm in situ. To free the frozen material from ice, the pressure in the apparatus was reduced and the main valve V_1 closed (see Figure 2). The valve V_1 served, that is, to permit the rapid evacuation of the system. The needle valve V_3, correctly set, served to prevent the loss of ice from the reservoir R (to the trap T) from interfering with the generation of the desired water vapor pressure.* Ice reservoir temperatures were selected to permit the reduction of the water activities in the samples to 0.5, 0.6, 0.7 and 0.8 at -40, -30, -20, and -10°C., respectively.**

Desorption isotherms were constructed from measurements accumulated when the samples were subjected to further, stepwise reductions in a_w. Resorption isotherms were derived from weights obtained on subsequent stepwise increase in a_w to final values of 0.6, 0.7, 0.8, and 0.9 at -40, -30, -20 and -10°C.** The balance recorded the completion of each step, indicating in addition to the final reading, the rate of approach to constant weight.

At 2° and 22°C. Isotherms were constructed from the measured weights of the samples exposed, in evacuated desiccators, to water vapor in contact with aqueous solutions of sulphuric acid /15/. Teflon-coated stirring bars, inserted beforehand, permitted the stirring of the sulphuric acid, in vacuo, for brief periods each day. Equilibria were achieved (i) in small enclosures maintained at 2 ± 0.05°C. in a room kept at 2 ± 0.5°C., (ii) in a room maintained at 21.5 ± 0.3°C. Desorption was effected directly from 10-ml. volumes of 20% aqueous PVP. Resorption isotherms were plotted from the weights of samples prepared from 20% aqueous PVP by freezing and freeze-drying to a very low water content in conventional shelf-type equipment. Water activities were, in each case, determined precisely from the densities of the sulphuric acid solutions after the completion of water transfer.

---

* The method just described differs from those in which manometric devices were used in that the working range was not determined by the lower working limit of the manometric controller. No difficulties arose in establishing water vapor pressures in the range 2 millitorr to 2 torr.

** The thermodynamic activities of ice are 0.68, 0.75, 0.82, and 0.91 at -40, -30, -20, and -10°C., respectively, when the vapor pressure of ice is expressed as a fraction of that of liquid water. Reasons for the choice of this form of expression are given in the Discussion.

Differential   Thermal Analysis

   Glass transition temperatures were determined from the rather
sudden changes in C_p with temperature characteristic of the glass
transition. The procedure, described in greater detail elsewhere,
for example, by Rasmussen and MacKenzie /16,17/, entailed the
cooling and the controlled rewarming of 0.005 to 0.010 g. of aqueous PVP at about 3 deg. C./min. The difference, ΔT, between the
temperature of the sample and that of a suitable reference was obtained as a function of the temperature, T, of the reference with
a dc amplifier/ X - Y recorder set-up. Glass transition temperatures were assigned to the samples by visual inspection of the
thermograms, the temperature at the point of inflection (that is,
where $d^2 \Delta T/dT^2$ = 0) being in each case chosen for the purpose.

RESULTS

   Typical weight/time information obtained with the equipment
illustrated in Figure 2 may be seen in Figure 3. Equilibria were,
generally, achieved in times ranging from several hours to one or
two days, depending on the sample temperature and the water vapor
pressure. Almost without exception, resorption was completed more
rapidly than desorption through the same step in water activity
at the same temperature. Equilibria were established in the larger samples in the desiccators at 2 to 22°C. in times ranging from
one to two weeks, in general. Weights of water retained or re-
gained by the PVP were determined from the weight/time recordings
or the weights of individual samples on the basis of dry sample
weights obtained afterwards at 22°C. Weighings were accurate to
±0.02 mg. at subzero temperatures and to ± 0.2 mg. at 2 and 22°C.

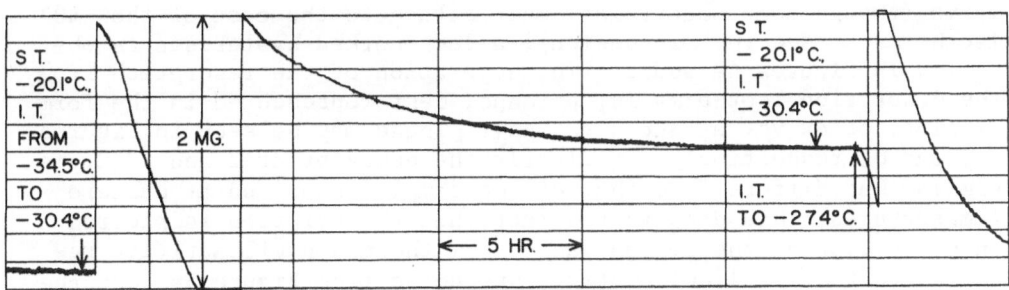

Figure 3. Typical weight/time trace obtained during the sorption
measurements at sub-zero temperatures. The recording illustrates
the changes in sample weight subsequent to the attainment of each
of two consecutive equilibrium values. The details are indicated.

Isotherms in which the weight of water sorbed per unit dry
weight of PVP was plotted with reference to water activity (a_w of
liquid water = 1 at each temperature) are displayed in Figure 4,
subdivided according to temperature. Each curve describing desorption subsequent to the removal of ice from the sample below
0°C. has been drawn to indicate its presumed origin on the curve
describing the temperature dependence of the water content of the
PVP in the presence of ice. The latter function, determined independently by MacKenzie /18/, has, for the purpose at hand, been
expressed in the form: weight of water per unit dry weight of PVP
vs. $(a_w)_{ice}$ where, as we noted, $(a_w)_{water}$ = 1. The same information is reproduced in the more conventional form (w/w composition
vs. temperature) in Figure 18.

One recognizes in Figure 4 the binding of very considerable
quantities of water and, at sub-zero temperatures, the retention
of even larger quantities in the presence of ice. One notes that
in every case the water content varies smoothly with a_w. The desorption isotherms gradually acquire a sigmoid character with increasing temperature; the resorption isotherms exhibit sigmoid
characteristics at each temperature, most obviously, perhaps, at
-10°C. Both desorption and resorption isotherms are lowered with
increase in temperature. Most remarkably, desorption isotherms
were found not to fall to zero water content at zero water activity but to low finite values that proved slightly greater the lower
the temperature. Water was, that is, retained during desorption
at 0 a_w at subzero temperatures that was removed at 0 a_w at the
conclusion of the desorption/resorption cycle when the specimen
was raised to 22°.

A very marked desorption/resorption hysteresis observed in
Figures 4a, b, c and d appears to terminate in the vicinity of the
intersection of the desorption isotherm with the freezing point
curve (though some uncertainty must attach to the case of the -40°
isotherms — see the Discussion); a less marked hysteresis is observed in Figures 4e and f. The separation of the resorption from
the desorption isotherms is, perhaps, best represented in the form
of plots of Δx vs. a_w and x vs. Δa_w. These may be seen in Figures
5a, and b, respectively. Clearly the behavior at 2 and 22°C. is
similar but differs from that of the samples examined at subzero
temperatures, this despite the fact that the sorption is plotted
in every case on the common basis of thermodynamic activity referred to that of liquid water. One notes from Figure 5a that the
area under the curve decreases from -40 to -30, increases to -20
and again to -10, after which it decreases abruptly to much lower
values. Clearly there is no single trend in hysteresis with temperature. Similar conclusions may be drawn from Figure 5b which,
however, shows more clearly the water contents between which hysteresis extends.

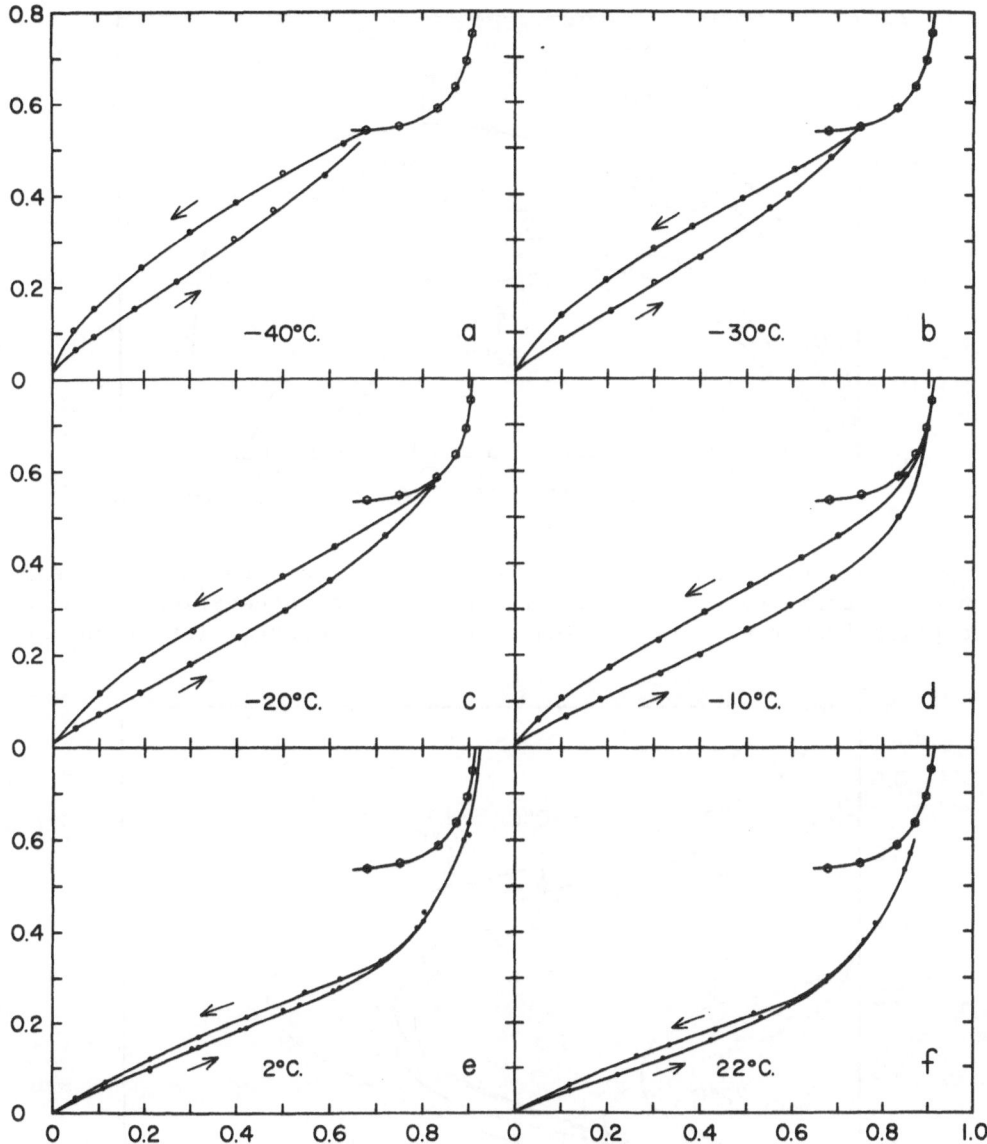

Figure 4 (a,b,c,d). Sorption isotherms obtained from frozen dilute
aqueous PVP by consecutive exposure to decreasing water activities
and, afterwards, to increasing water activities.   Figure 4 (e,f).
Sorption isotherms obtained from dilute aqueous PVP and (in the case
of Figure 4e) from freeze-dried PVP (see text).   The freezing point
curve determined by MacKenzie /18/ has been inserted in each of the
six figures. Abscissae: water activity; ordinates: g. H_2O/g. PVP.

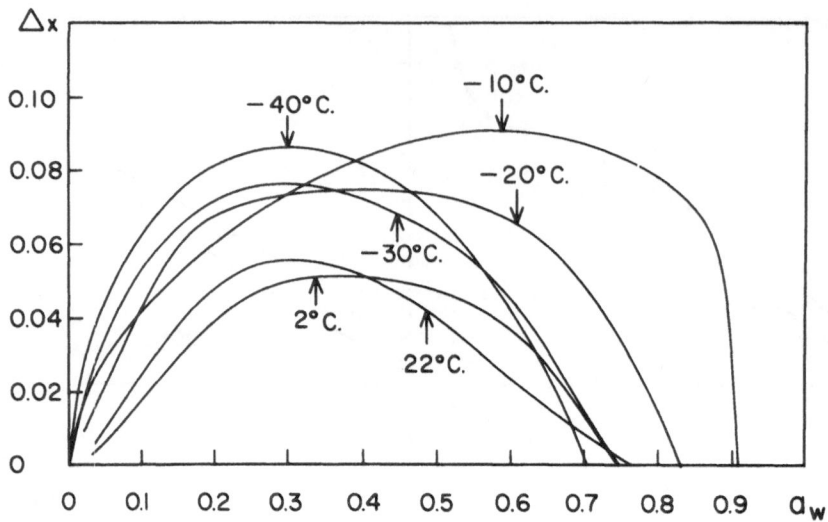

Figure 5a. Plots showing the dependence of the sorption hysteresis on water activity. Δx represents the vertical separation of the desorption from the corresponding resorption isotherm (see Figure 4).

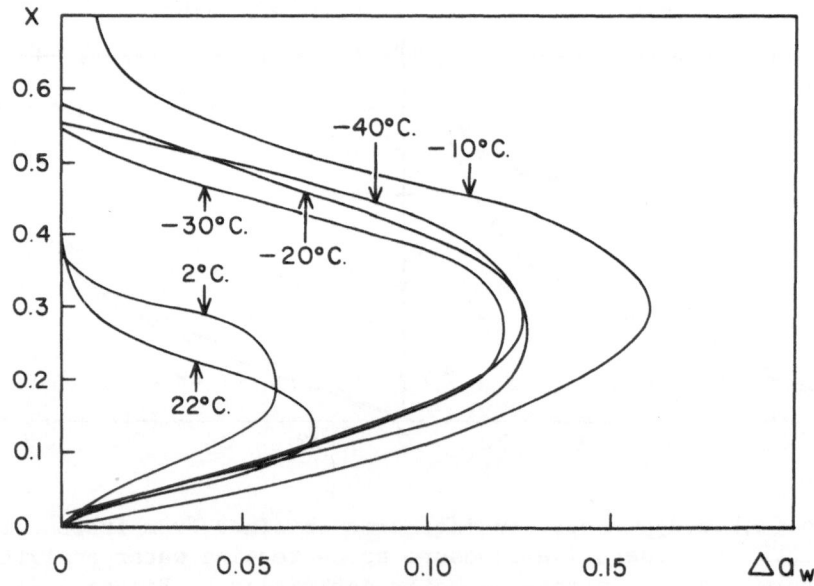

Figure 5b. Plots showing the dependence of the sorption hysteresis on water content. Δa denotes the horizontal separation of the desorption from the corresponding resorption isotherm (see Figure 4).

Thermograms obtained on differential thermal analysis of 55, 65, and 75% weight/weight aqueous PVP are reproduced in Figure 6. The 65 and 75% solutions, being too concentrated to freeze at all, exhibited only the transition from glassy to viscous states. The 55% solution, freezing readily when subjected to slow cooling, was rendered vitreous by rapid cooling and exhibited, on being warmed, first a glass transition, second a spontaneous freezing of water, third a melting of ice, as indicated in the figure. Corresponding thermograms, not shown, were obtained from 50, 60 and 70% solutions. The 50 and 60% samples behaved in much the same manner as the 55% sample; the 70% solution behaved similarly to the 65 and 75% specimens. The points of inflection to which we referred in the previous section are, it is seen, readily determined from the recording to ± 0.01 mv, i.e., to ± 0.3 deg. C.

Glass transition temperatures determined from the thermograms were plotted with reference to composition. The resulting dependence of the T_g on the water content is shown in Figure 7. The curve has been drawn to pass through -137 and 86°C. in accordance with published values for the T_g of H_2O /17/ and anhydrous PVP /19/. See also Figure 18. One notes that the values obtained were all derived from measurements at warming rates of about 3 deg. C./min. The dependence of the glass temperatures on the warming rate was not determined. One notes, in particular, that the glass temperature increased by 70 deg. C. when the w/w ratio: water/PVP was decreased from 1/1 to 1/3.

The better to visualize the separation of the desorption isotherms, one from another, and the temperature dependence of the water activity where the material assumes a glassy state on isothermal dehydration, w/w compositions interpolated from Figure 7 have been transformed and combined with information from Figure 4 to yield plots reproduced in Figure 8. Correspondingly, the same interpolated w/w compositions have been combined with resorption isotherms from Figure 4 to yield Figure 9 to which has been added an isotherm at 25° constructed from measurements reported by Dole and Faller /20/.

One sees at once that sorption is indeed strongly dependent on temperature. The dependence appears to exhibit a discontinuity (between -10 and 2°C.) in desorption (Figure 8) but to be continuous, that is, uniformly graded from -40 to 22°C., with respect to resorption (Figure 9). It appears also from Figures 8 and 9 that there is, generally, a tendency to a greater separation of the isotherms, one from another, in resorption; the intersection of the 22° isotherm with the 2° curve between .7 and .75 a_w provides the solitary exception. One notes that the gradients of the T_g plots are rather different, depending on the direction in which the sorption is proceeding, though both T_g plots must curve quite sharply to pass through the point: 0 bound water, 0 water activity

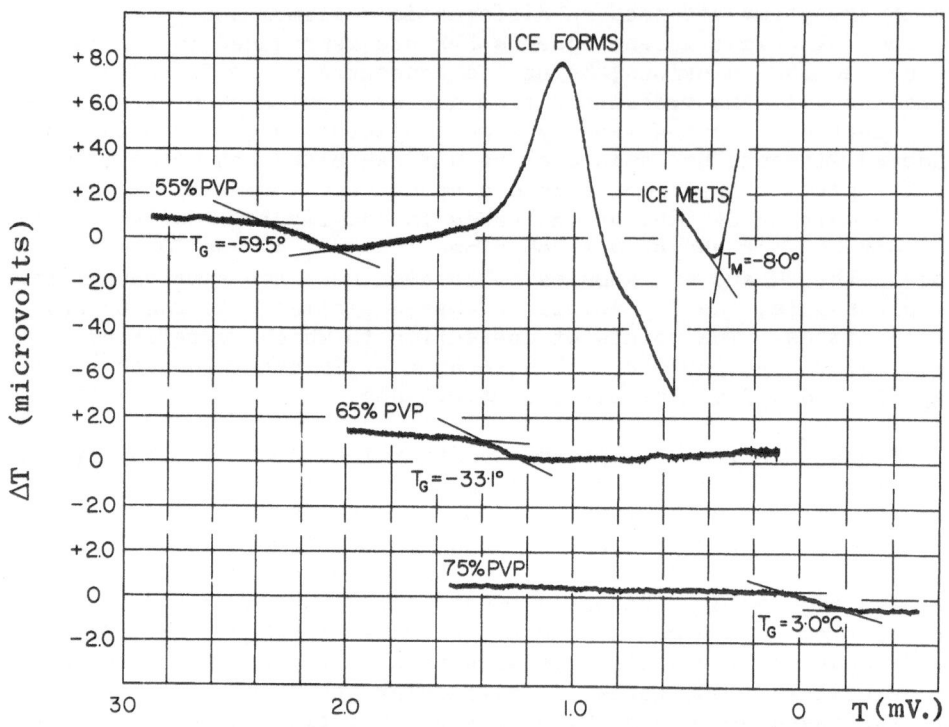

Figure 6.   Thermograms by differential thermal analysis of aq. PVP.

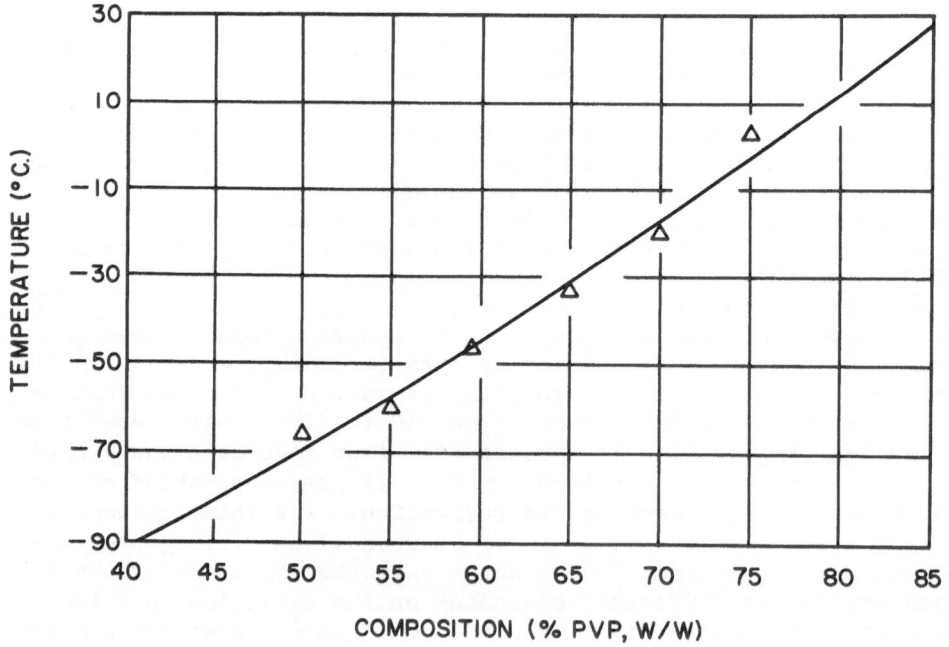

Figure 7.   Dependence of the glass temperature on the water content.

(at which point they would, one suspects, have to become tangents to the respective 86°C. isotherms).

The relationship of the subzero desorption isotherms to the freezing point curve is apparent from Figure 8. Freezing and freeze-drying necessarily involve the passage of a sample through one or another series of states described by the descent of the freezing point curve and, at constant freeze-drying temperature (no thermal gradients in the sample), the further descent of one of the desorption isotherms. One notes also that desorption isotherms proceed from the freezing point curve only where frozen samples are subjected to desorption. Desorption from suitably dilute aqueous PVP supercooled to -30 or, for example, -40°C. could have occurred from points far above the freezing point curve in the absence of any ice in the sample.

The freezing point curve has been omitted from Figure 9 in as much as resorption, in the absence of the formation of ice in an experimental set-up, might be expected to proceed at subzero temperatures regardless of the existence of the freezing point curve. In practice, however, attempts to create vapor phase water activities higher than those of ice resulted in the deposition of ice on the walls of the glass specimen chamber in the neighborhood of the specimen.*

Since errors in water activity (resulting from errors in the control of ΔT) and in measured sample weights are estimated to fall within the size of the circles denoting the data points (see Fig. 4a, b, c, d, e, f), the temperature dependence of the sorption is clearly demonstrated. Modes of passage into the glassy state by desorption or by decrease in temperature and of passage into the viscous state by resorption or by increase in temperature are, similarly, clearly described by the T_g plots (the broken lines dividing Figures 8 and 9 each into two regions).

For the sake of subsequent discussion the sorption data were also plotted with reference to the thermodynamic activity of ice; $(a_w)_{ice} = 1$ at all temperatures. The resultant plots are reproduced in Figures 10 and 11. Examined in this form, the data indicate, in general, (i) a slight but definite temperature dependence at lower water contents and water activities, (ii) a marked temperature dependence at higher water activities, (iii) the clear independence of the desorption isotherm at 2°C. from the remaining desorption isotherms (activities of water vapor at 2°C. were obtained by referring water vapor pressures to that of ice at 2°C., an extrapolated value for the latter pressure being assumed).

---

* Neither were we able to deposit ice in the resorbed specimen, it being impossible in practice to cause a sample to traverse a path somewhat akin to the reverse of the freeze-drying procedure.

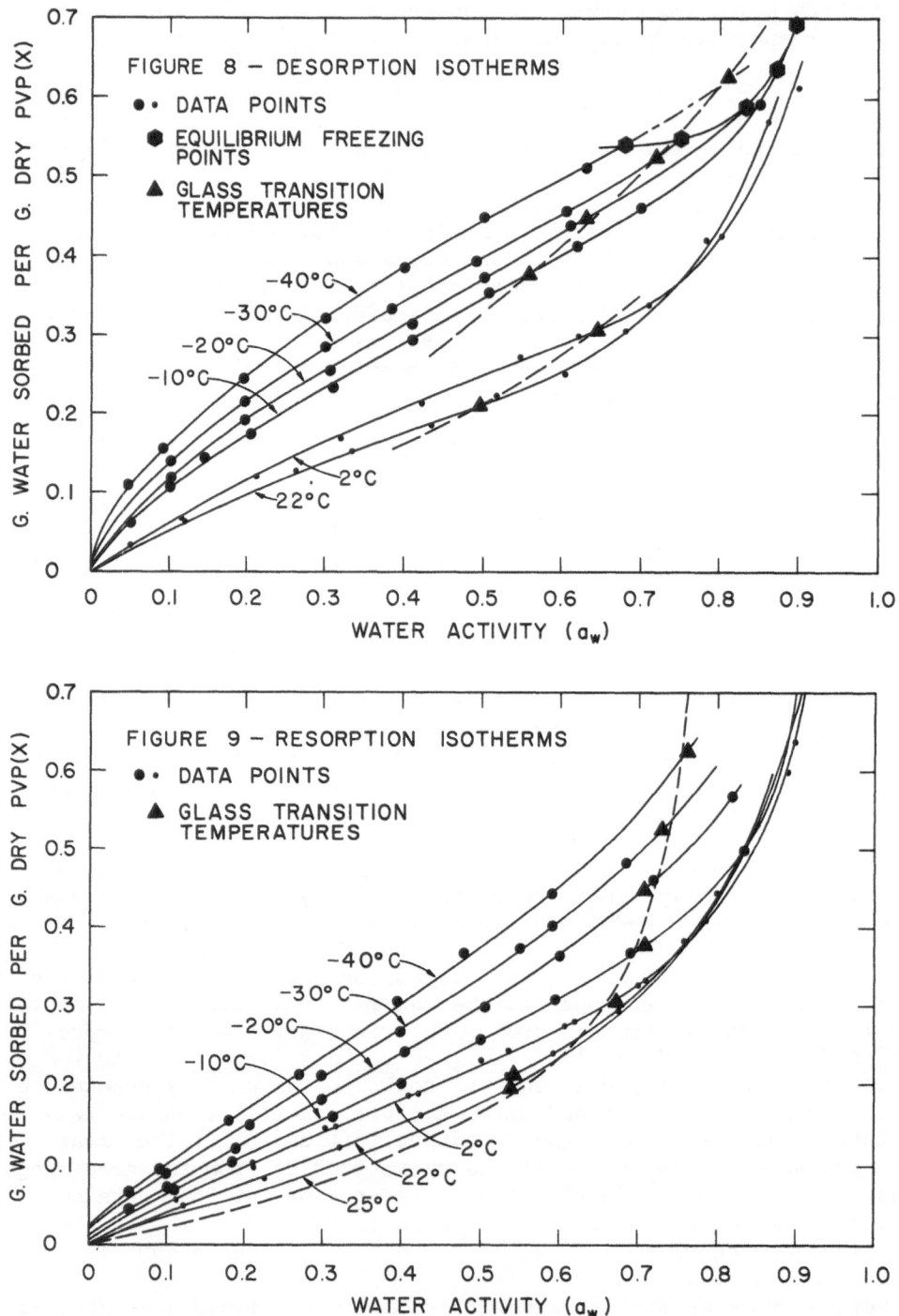

Figures 8 and 9. Sorption data replotted with reference to a$_{water}$.

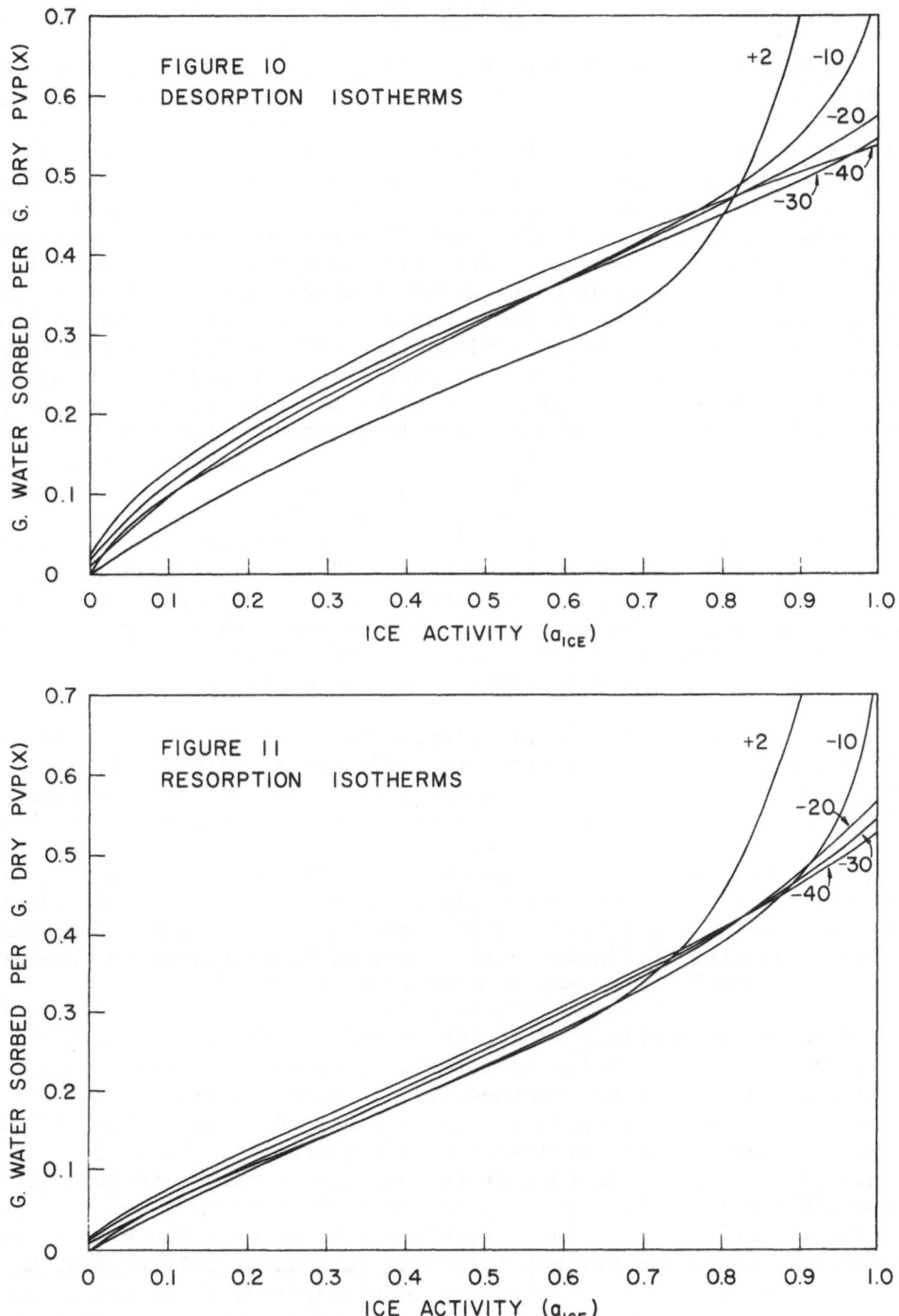

Figures 10 and 11.   Sorption data replotted with reference to a_{ice}.

DISCUSSION

An examination of Figures 8, 9, 10 and 11 reveals the very considerable effect of the manner of plotting on the shapes and inter-
relationships of the sorption isotherms. When the sorption data are
graphed with reference to the activity of liquid water the isotherms
exhibit typical forms, tending in each case to infinite sorption at
unit a_w. They are widely spaced, indicating, from the direction of
the spacing, a large, positive heat of sorption, that is, a strong
binding of water to polymer. The relationship of the desorption
isotherm to the freezing point curve inserted in Figure 8 is readily
recognized, dehydration of the dissolved polymer by freezing being
quite clearly the necessary prelude to further dehydration by freeze-
drying. The regularity of the spacing of the resorption isotherms
(Figure 9) suggests adsorption occurs by the same mechanism throughout the range -40 to 22°C. and that it proceeds without regard (i)
to the freezing point of water, (ii) to the way the sample was prepared, since isotherms at 2°C. obtained from freeze-dried and from
regularly dried PVP fell one on the other. The interruption in the
spacing of the desorption isotherms between -10 and 2°, and the sud-
denly decreased hysteresis (Figures 5a, 5b, and 8) suggest, on the
other hand, that desorption proceeds in one of two ways, depending,
apparently, not on the temperature but on the method of preparation.
Only in the crossing of the 2 and the 22°C. plots in each case does
there appear a need for a further, special explanation.

When the sorption data are plotted with respect to the activity
of ice, the sorption isotherms are, in general, rather less informa-
tive. In neither case is the temperature dependence sufficiently
large and distinct to suggest the value of a thermodynamic analysis.
Nor is it possible to relate the isotherms to the function describing the quantities of water remaining in PVP in the presence of ice.
One notes, moreover, the seemingly complex series of intersections
at higher water activities. From a practical viewpoint, then, the
plots reproduced in Figures 8 and 9 are to be preferred to those of
Figures 10 and 11. (See editor's note on page 172.)

From the theoretical viewpoint, it would appear more logical
to define water activities in terms of a condensed state to which
sorbed water could tend spontaneously at each temperature, with increasing a_w. In as much as it can be supercooled almost to -40°C.
/21, 22/, liquid water satisfies this requirement. There is, in
contrast, no way to maintain or to form ice at temperatures higher
than 0°C.; neither is there any reason to believe that water sorbed
to PVP can assume repeating structures exhibiting a truly hexagonal
symmetry. Where ice has not formed in an aqueous polymer system,
that is, where nucleation of I_h has not occurred, there can be no
sudden change in the nature of the ordering of water present with
temperature. Where water activities are referred to liquid water
there can, that is, be no discontinuity in the temperature dependence of the isotherms in the neighborhood of 0°C. (see Figure 9).

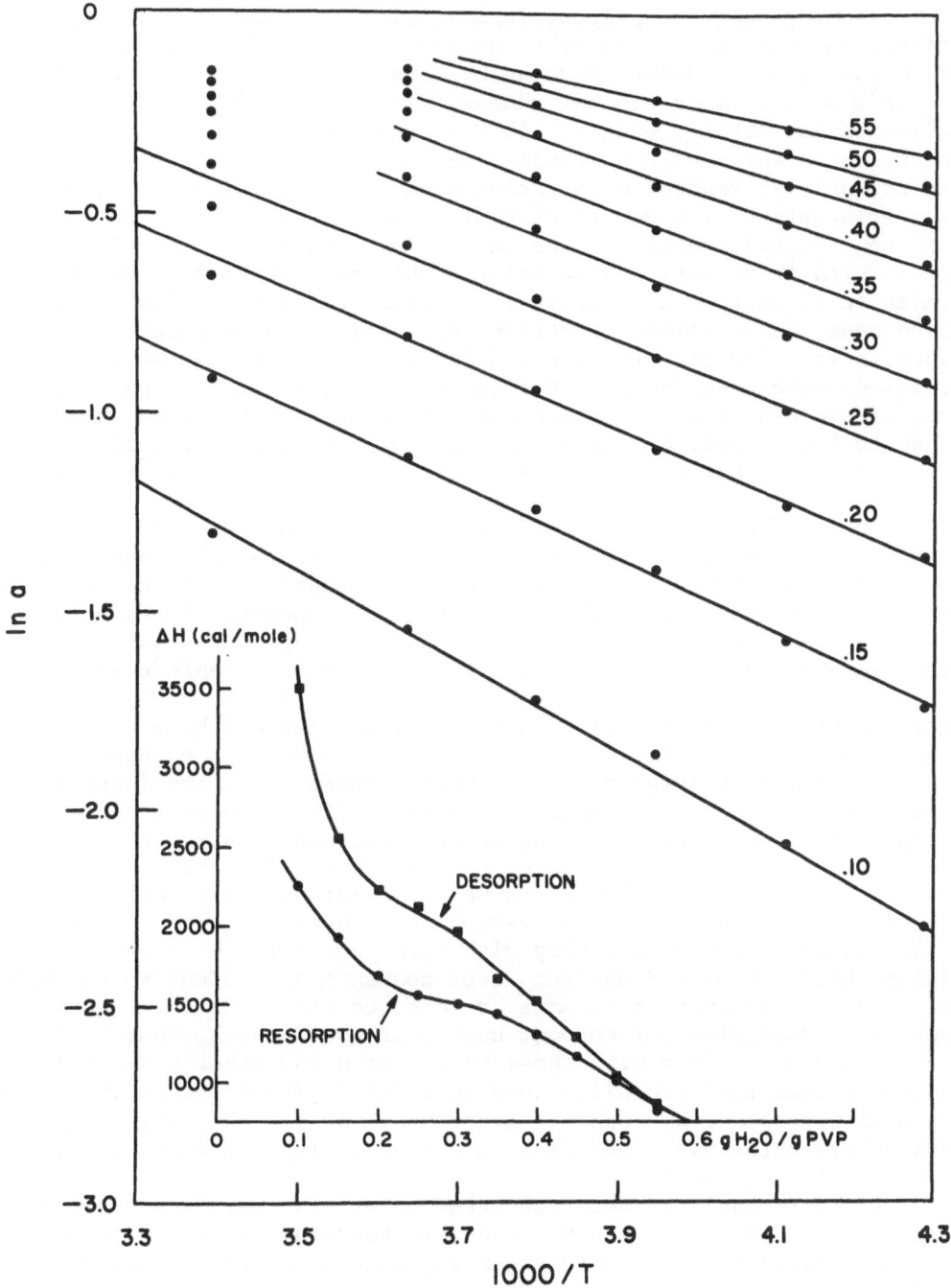

Figure 12. Ln a vs. 1/T plots obtained at various water contents from the resorption isotherms. Inset: dependence of the heat of sorption on the water content and direction of the sorption process.

Knowing that the polymer is miscible with water in all proportions, we might expect water to desorb smoothly with decreasing a_w
from supercooled, dilute aqueous PVP and to resorb with increasing
a_w in a more or less similar manner, yielding an infinitely dilute
solution at unit a_w, even at -40°C. Accordingly, we could not expect ice to serve as a suitable reference state unless water sorbed
to the polymer tended, with increase in water activity, to acquire
the structure of ice and to accumulate in the polymer in the form
of ice. Clearly, such an expectation requires PVP to be soluble in
ice, which it is not, and sorption to be indeterminate rather than
infinite at unit water activity. In practice, PVP retains, as we
have seen, well-defined quantities of water in the presence of ice,
down to very low temperatures and is, itself, as far as we can determine, insoluble in ice. For the reasons just listed the remainder of the discussion is presented with reference to water rather
than to ice.* Determinations of heats of sorption, and analyses
according to Brunauer, Emmett and Teller, and Bradley are reported.

Enthalpies were determined from plots of ln a vs. 1/T, 'a' denoting the water activity, at temperature T, of the PVP containing
a given amount of water, changes in heat content, ΔH, being obtained
from the gradient, $\Delta H/R$, where R is the gas constant. Plots obtained
from the resorption isotherms (Figure 9) are reproduced in Figure 12;
the heats derived from the best straight lines are reproduced in the
inset figure, plotted with reference to water content. While the
corresponding ln a, 1/T plots obtained from Figure 8 have not been
inserted in Figure 12, for lack of space, the derived changes in
heat content on desorption have been incorporated in the inset figure. It is apparent (i) that the changes in heat content on desorption exceed the changes on resorption by several hundred calories
per mole, except at higher water contents where the differences dis-
appear, (ii) that the changes in heat content exceed in each case
the latent heat of fusion at lower water contents, (iii) that the
heats decrease rather smoothly with water content, the plots exhibiting shoulders extending from water contents equivalent to one water
molecule per monomer unit, more or less, to water contents equal to
two water molecules per monomer unit. It would appear that the findings are more in line with those on proteins and other hydrophilic
polymers than they are with those obtained from water-insoluble polymers incorporating hydrophilic groupings, PAN and PMM, for example
/7/, where enthalpies increased numerically with the water sorbed.

B. E. T. analyses were conducted on all the sorption isotherms
displayed in Figures 8 and 9. They are reproduced, in the form of
plots of a/x(1 - a) vs. a (where x represents the water sorbed by

---

*It should, for the same reasons, have become clear that no
useful purpose would be served by discussing sub-zero data with reference to $(a)_{ice}$ and data from determinations at higher temperatures
with reference to a_w.

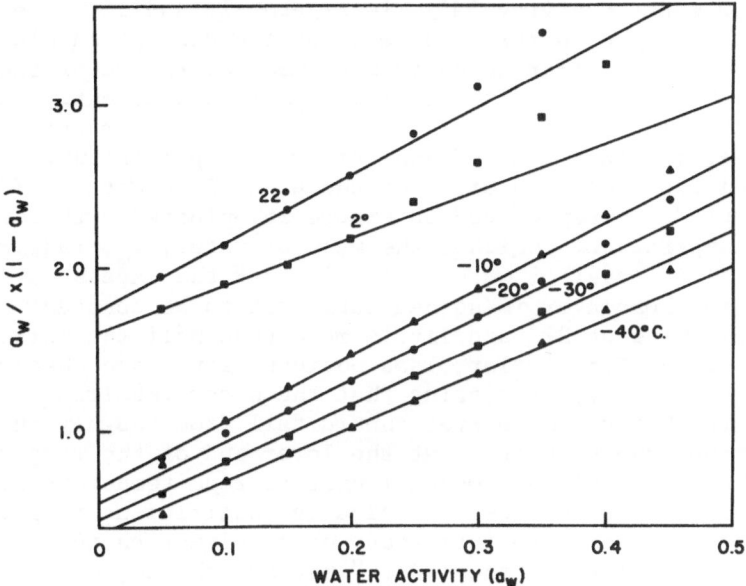

Figure 13. B. E. T. plots derived from the desorption isotherms.

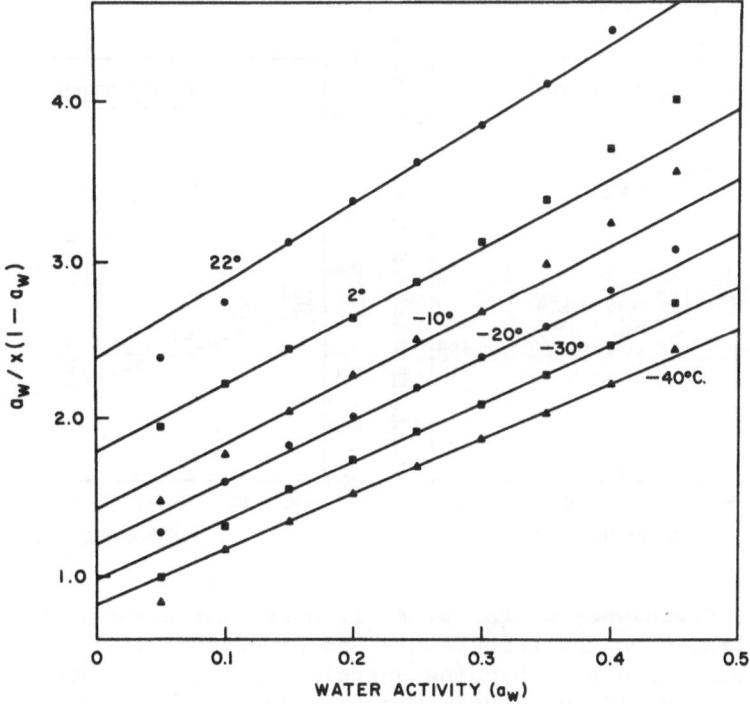

Figure 14. B. E. T. plots derived from the resorption isotherms.

the PVP at a water activity 'a'), in Figures 13 and 14. One notes
the near linearity of portions of each of the desorption plots in
Figure 13 and the similar appearance of most of the resorption plots
in Figure 14. One notes also the rather more consistently sigmoid
nature of the plots in Figure 14. Gradients and intercepts taken
from the best straight lines drawn through the points in Figures 13
and 14 permitted the calculation of the B. E. T. constants /3/.
The "constants" are reproduced in Figure 15, plotted with reference
to sample temperature. Neither the mass of water, x_m, required to
complete a first layer of water molecules nor the excess energy, E_m,
with which the monolayer is sorbed turns out to be constant. The
values x_m exhibits at 22° are little more than half the corresponding values at -40°C.; similar, less obvious trends are observed in
E_m. It is especially interesting that the water required to complete the hypothetical monolayer should fall from roughly three molecules per two monomer units, at the lower end of the range examined,
to roughly one molecule per monomer unit at room temperature. Quantities of water of this order are clearly insufficient to permit a
coat of water molecules to form about each polymer chain, or to permit extensive sharing of "monolayers" between chains, or even to
allow the formation of closed structures encasing only the hydrophilic = N - (C = O) - groups. It would appear, rather, that
we must interpret the term "monolayer" in the light of other findings.

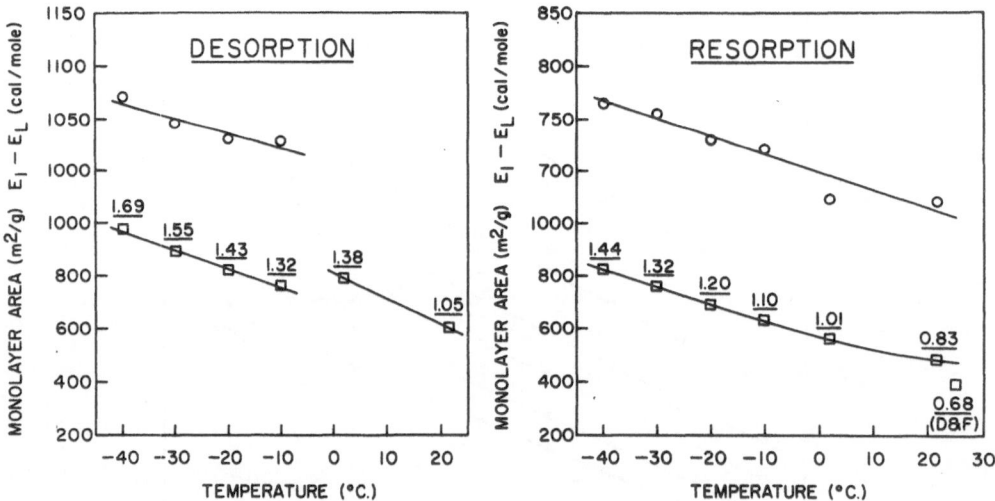

Figure 15. Dependence of the B. E. T. monolayer areas and excess
binding energies on the temperature and direction of the sorption
process. Circles denote binding energies; squares represent monolayer areas; underlined numbers denote moles water per monomer unit.

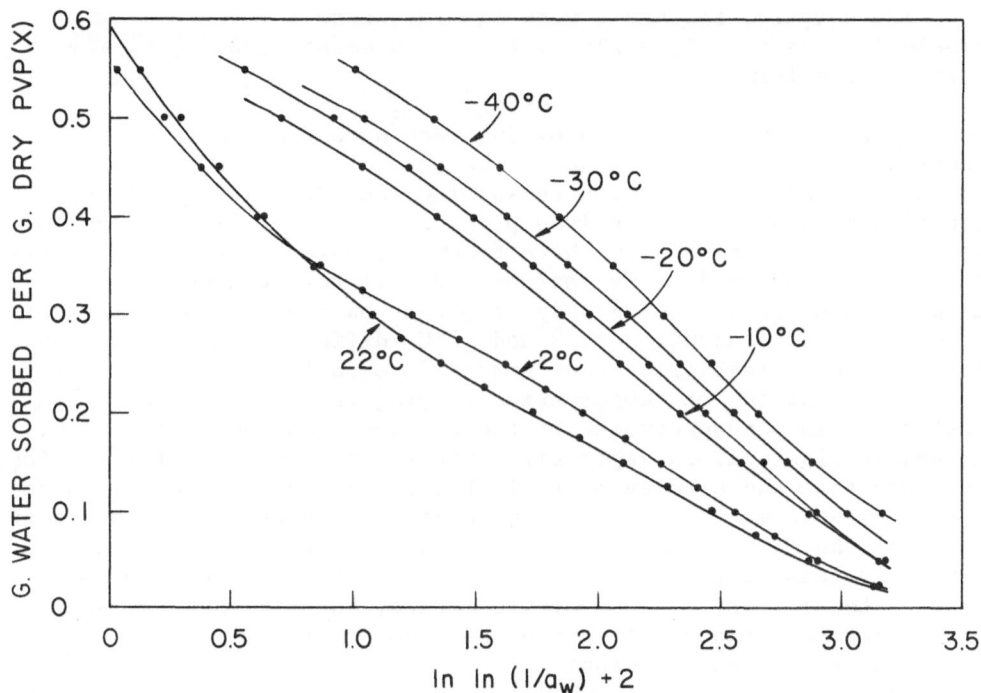

Figure 16.   Bradley plots derived from the desorption isotherms.

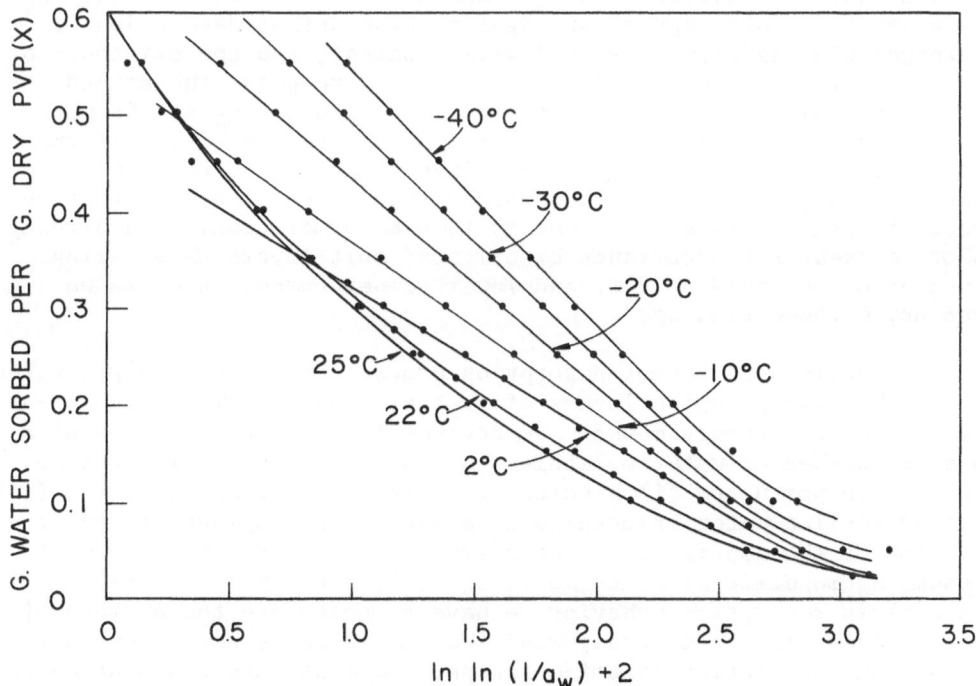

Figure 17.   Bradley plots derived from the resorption isotherms.

The sorption isotherms were examined also for conformity to
Bradley's equation /4,5/ which, for the present purpose, we may
write in the form:

$$- \log a = K_1 K_2^x + K_3 ,$$

where a and x retain their previous meaning and K_1, K_2, and K_3 are
constants. To test experimental measurements, values of x, the
quantity sorbed, are plotted versus log log 1/a. Where Bradley's
equation is obeyed, the derived plots are linear. Figures 16 and
17 depict plots in which we have taken our values for x from Figures
8 and 9, respectively. We can conclude (i) that the behavior during
desorption differs considerably from that on resorption, (ii) that
the process of desorption at 2 and 22°C. differs from that at the
lower temperatures, (iii) that Bradley's equation is obeyed only
during resorption, at temperatures ranging from -40 to 2°C. The
difference in the appearance of the 2° curve and the 22 and 25°C.
curves in Figure 17 is especially interesting. Equally interesting,
it seems that the Bradley plots in Figure 17 are linear only where
water contents exceed, roughly speaking, the quantities required to
complete the B. E. T. monolayer. At lower water contents, the Bradley plots tend to curve rather sharply despite the manner of plotting -- the latter, because of its method, ensures that the shapes
of the curves are not affected by zero point errors in the determination of the water contents.

What, then, can we say concerning the disposition of the water
that is not in conflict with the results of the foregoing analyses?
Ln a vs. 1/T plots suggest stronger binding during desorption, a
stronger binding with decreased water content, and the existence of
a number of equivalent sites accommodating, roughly, the second
water molecule per monomer unit. B. E. T. plots suggest the monolayer concept applies to both desorption and resorption, that the
monolayer area varies smoothly with the temperature, that the area
involved is greater during desorption than resorption, and that the
water is bound with greater energy in the former case. The Bradley
plots suggest the occurrence of oriented multilayers only during
resorption and only at 2°C. and lower temperatures. How can we
reconcile these findings?

It would appear that desorption proceeds via a series of states
in which water occupies layers of greater extent, being less oriented, and more strongly bound. To account for the accomodation of a
greater number of water molecules and for the lack of orientation,
we have to postulate (i) binding of water by hydrophilic sites, (ii)
clathrate-like water structures necessary, one imagines, to the retention of the configuration of polymer chains. Further, we must
assume hydrophilic sites to be more nearly available from all sides.
To explain resorption behavior we have to postulate the absence of
generalized clathrate structures. We must also assume a tendency
toward an association of amide groups. We must assume hydration to
proceed during resorption by the formation of consecutive shells of

water molecules aligned with reference to hydrophilic groups or clusters of groups; that is, we must assume water to be incorporated during rehydration into a structure in which it is more highly localized than that from which it was lost during resorption. A decreasing "monolayer area" with increasing temperature parallels, most probably, an increasing freedom of internal movement in the polymer and an increasing tendency to the elimination of isolated hydrophilic groups.

The loss of water from the structures arising during desorption introduces, presumably, strains accounting for the higher ΔH. The filling of different sites on resorption reduces these strains, explaining the lesser ΔH. It would seem, then, that desorption annihilates an extensive water - polymer "interface" which, to be re-created by a reversible resorption, would require an event equivalent to a spontaneous 2-dimensional nucleation. Presumably the nucleation is energetically improbable, requiring, as it must, the cooperative interaction of polymer chain segments at low water content. At higher water contents or higher temperatures, where such a nucleation might have occurred, the polymer chains have already become so mobile that other, less structurally restricting forms of interaction predominate.

The notion of the persistence of the coherent interface on desorption requires that the water molecules attached to the hydrophilic parts of the polymer be directly associated, one with another, or that they be bridged by water which must, of necessity, make contact with apolar surfaces. A coherent monolayer must, that is, be composed of "true" monolayer and second (and third) layers of water, judged with reference to the hydrophilic constituents. Actual arrangements of the water are, perhaps, best discussed from this point of view, in the knowledge that the number of primary water molecules has been found to vary from 3 per 2 amide groups at −40° to 2 water molecules per 3 such groups at 25°C. (Figure 15).

Returning to Figures 8 and 9, and to questions relating to the glass transition, we note that the spacing of the sub-zero isotherms is roughly constant, though it was only at −40 that freeze-drying took place below the glass temperature of the polymer phase adjacent to the subliming ice. Only at −40 and −30°C., moreover, did freeze-drying proceed with retention of the solute matrix structure; collapsed materials were obtained at −20 and −10°C.* It is interesting that neither physical event introduced a discontinuity into the temperature dependence of the sorption process. It is particularly interesting that in the case of the resorption process the mode of preparation should not have introduced a discontinuity in the tem-

---

*The collapse temperature, the temperature above which freeze-drying proceeds with structural collapse of the solute matrix, was determined previously to be −23 ± 1°C. for PVP /23/.

perature dependence between -10 and 2°C. (and that isotherms obtained at 2°C. from freeze-dried and from regularly dried polymer should coincide). Since a discontinuity is observed in the desorption isotherms between -10 and 2°C. we must assume that the means employed to effect the preliminary dehydration determines the further course of the desorption. Very possibly, the freezing prior to freeze-drying, concentrating the PVP to 57 to 65%, depending on the extent of the rewarming to the freeze-drying temperature, restricted the polymer to states in which, for example, pyrrolidone rings were less frequently clustered than those in the PVP dried from dilute solution in the desiccators.*

We come now to the effect of the structure of water on the sorption process, having dealt with the effect of the polymer. To explain the properties of liquid water at lower temperatures, it has been supposed that the liquid exhibits a greater "structuring" the lower the temperature, acquiring, in the process, more of an ice-like character /24, 25, 26, 27, 28/. It is likely, therefore, that water molecules farthest from the polymer chains in very dilute aqueous PVP demonstrate a more extensive structuring, the lower the temperature. It is likely that the difference between the partial molar free energy of this, the least strongly bound water, and that of ice decreases correspondingly. While actual measurements are not available it appears that p_{ice}/p_{water} decreases with temperature less rapidly than might have been supposed from the Clapeyron equation and the measured latent heats at 0°C., the more so the lower the temperature.

If this be the case it would seem that Figures 8 and 9 were better plotted to allow for the modified dependence of the ratio of the vapor pressure of ice to that of water on the temperature. We should, that is, have decreased the lateral spacing of the isotherms with the temperature. The linearity of the B. E. T. and the Bradley plots, when it was obtained, would not have been affected; small changes in the ln a, 1/T plots would have arisen, requiring an explanation, but might have passed undetected.

The question is, in any case, unresolved in the absence of a better knowledge of the structure of water below 0°C. Plotting isotherms on common coordinates with reference to values of a_{ice}/a_{water} derived on the basis of the Clapeyron equation, we would, it appears, obtain from isotherms at -60, -80, -100, etc. (i.e., from our ln a, 1/T plots), an indication of a redissolution of ice, with cooling, at very low temperatures. Riedel examined this question with reference to frozen, heat-denatured muscle tissue /29/.

─────────────────────

*In this last respect, it is very possible that the glass temperatures we determined on highly concentrated aqueous PVP prepared at 60 to 80°C. are only approximately applicable to the variously hydrated freeze-dried states of corresponding water content.

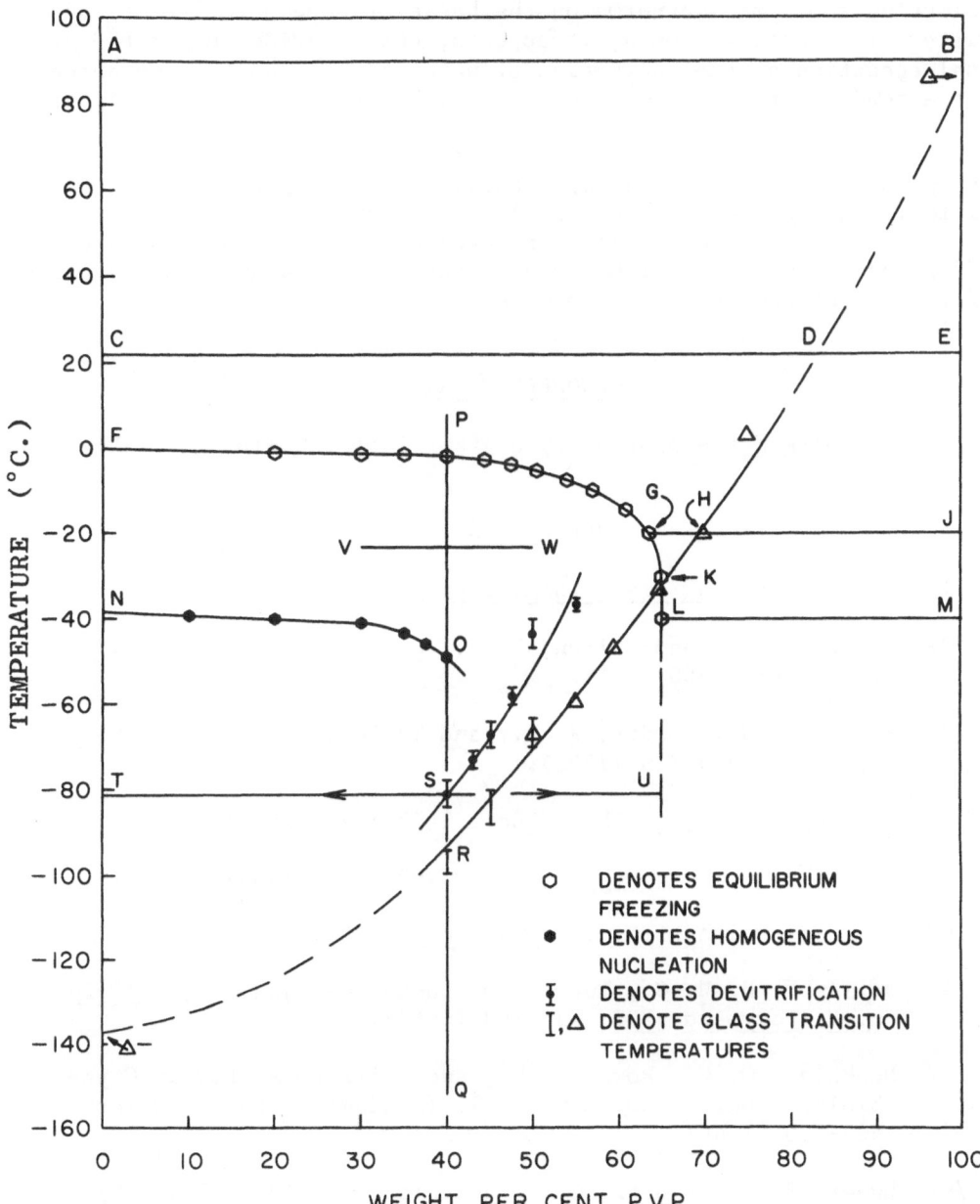

Figure 18. The system: water – PVP. Dilute aqueous PVP enters the glassy state on dehydration (i) at higher temperatures, e.g., at D, (ii) only during freeze-drying, e.g., at H, (iii) at lower temperatures, during freezing, e.g., at K. Ultra-rapid cooling from P to Q yields a glassy sample that (i) softens on warming at R, (ii) devitrifies at S, disproportionating to T and U, (iii) exhibits the migratory recrystallization of ice above the line VW /18,19,22,30/.

Plotting the same isotherms on the basis of a temperature dependent water structure, we could, it appears, obtain evidence for the gradual cessation of the conversion of water to ice and for the absence of a tendency to redissolution. The subject deserves investigation.

From yet another comparative point of view the course of the sorption processes can be represented without regard to water activities. This has been done, lastly, by combining, in Figure 18, the measured freezing points /18/, homogeneous nucleation temperatures /22/, and glass transition temperatures from this and other studies /19,30/. Figure 18 is, we hope, entirely self-explanatory.

ACKNOWLEDGEMENT

This work was supported by a grant from the NIH (HE-11386).

REFERENCES

/1/   Bull, H. B., J. Amer. Chem. Soc., 66, 1499-1507 (1944).

/2/   Hoover, S. R. and Mellon, E. F., J. Amer. Chem. Soc., 72, 2562-2566 (1950).

/3/   Brunauer, S., Emmett, P. H., and Teller, E., J. Amer. Chem. Soc., 60, 309-319 (1938).

/4/   Bradley, R. S., J. Chem. Soc., 1467-1474 (1936).

/5/   Bradley, R. S., J. Chem. Soc., 1799-1804 (1936).

/6/   Ling, G. N., Ann. N. Y. Acad. Sci., 125, 401-417 (1965).

/7/   Jellinek, H. H. G., Luh, M. D., and Nagarajan, V., Kolloid-Z. u. Z. Polymere, 232, 758-763 (1969).

/8/   Doebbler, G. F., Rowe, A. W., and Rinfret, A. P., in Cryobiology, ed. by Meryman, H. T. (Academic Press, New York) 407-450 (1966).

/9/   Luyet, B. and Rapatz, G., Cryobiology, 6, 425-482 (1970).

/10/ Mazur, P., Farrant, J., Leibo, S. P., and Chu, E. H. Y., Cryobiology, 6, 1-9 (1969).

/11/ Mazur, P., Leibo, S. P., Farrant, J., Chu, E. H. Y., Hanna, M. G., and Smith, L. H., in The Frozen Cell, ed. by Wolstenholme, G. E. W. and O'Connor, M. (J. A. Churchill, London) 69-85 (1970).

/12/ Greaves, R. I. N. and Davies, J. D., Ann. N. Y. Acad. Sci., 125, 548-558 (1965).

/13/ Luyet, B. and Rapatz, G., Biodynamica, 7, 337-345 (1957).

/14/ MacKenzie, A. P. and Luyet, B. J., Biodynamica, 9, 193-206 (1964).

/15/ Blake and Greenwalt, International Critical Tables, 3, 302 (McGraw-Hill, New York) (1928).

/16/ Rasmussen, D. H. and MacKenzie, A. P., Nature, 220, 1315-1317 (1968).

/17/ Rasmussen, D. H. and MacKenzie, A. P., J. Phys. Chem., 75, 967-973 (1971).

/18/ MacKenzie, A. P., Cryobiology, 8, 379 (abstract) (1971).

/19/ Sugiura, M. and Fujii, E., Kogyo Kagaku Zasshi, 66, 1228-1230 (1963).

/20/ Dole, M. and Faller, I. L., J. Amer. Chem. Soc., 72, 414-419 (1950).

/21/ Fletcher, N. H., The Chemical Physics of Ice, (Cambridge University Press, London) (1970).

/22/ Rasmussen, D. H. and MacKenzie, A. P., this volume.

/23/ MacKenzie, A. P., Bulletin of the Parenteral Drug Association, 20, 101-129 (1966).

/24/ Frank, H. S. and Quist, A. S., J. Chem. Phys., 34, 604-611 (1961).

/25/ Wada, G., Bull. Chem. Soc. Japan, 34, 955-962 (1961).

/26/ Némethy, G. and Scheraga, H. A., J. Chem. Phys., 36, 3382-3400 (1962).

/27/ Davis, C. M. and Litovitz, T. A., J. Chem. Phys., 42, 2563-2576 (1965).

/28/ Jhon, M. S., Grosh, J., Ree, T., and Eyring, H., J. Chem. Phys., 44, 1465-1472 (1966).

/29/ Riedel, L., Kältetechnik, 13, 122-128 (1961).

/30/ Luyet, B. and Rasmussen, D. H., Biodynamica, 10, 137-147 (1967).

EDITOR'S NOTE

The paper on "Freezing of Aqueous Polyvinylpyrrolidone Solutions" by H. H. G. Jellinek and S. Y. Fok, Koll. Zeitschr. und Zeitschr. für Polymere, <u>220</u>, 122, (1967), should be consulted in this connection. Their results were evaluated on the basis of ice activities which give a type III isotherm with zero thermal coefficient. Isotherms obtained from water sorption (see ref. 7 of the above paper) evaluated on the basis of water activities are also practically independent of temperature.

AUTHOR INDEX

Underlined numbers indicate complete references

SUBJECT INDEX